INVESTIGATING ECOLOGY

INVESTIGATING ECOLOGY

Elliott H. Blaustein
and
Rose Blaustein

**Arco Publishing Company, Inc.
New York**

Published by Arco Publishing Company
219 Park Avenue South, New York, N.Y. 10003

Copyright © 1978 by Elliott H. Blaustein and Rose Blaustein

All rights reserved. No part of this book may be
reproduced, by any means, without permission in
writing from the publisher, except by a reviewer
who wishes to quote brief excerpts in connection
with a review in a magazine or newspaper.

Library of Congress Cataloging in Publication Data

Blaustein, Elliott H
   Investigating ecology.

   Bibliography, p. 135
   Includes index.
   1. Ecology. 2. Ecology—Technique. I. Blaustein,
Rose, joint author. II. Title.
QH541.B55   574.5′028   76-55356
ISBN 0-668-04440-3 (Library Edition)
ISBN 0-668-03853-5 (Paper Edition)

Printed in the United States of America

To Elizabeth Guthrie
in recognition of her devotion to
environmental education

ACKNOWLEDGMENTS

by the authors
To Louis Sackman, whose skillful, creative
editing greatly enhances the worth of this book

by Arco
To the municipal government of Oakland, New
Jersey, for their assistance in locating and obtaining
real estate maps and maps of local flood plains
To the St. Regis Company for permission to
take photographs on their property in
West Nyack, New York

Botanical drawings by Elliott H. Blaustein
Other drawings by Jim Backes
and Louis Sackman
Photographs by Alan Jay Federow
Models for these photographs are
Valerie Turner, Ronald Adolf,
Conné Kershner, and James Stevens

CONTENTS

	INTRODUCTION	11
1.	HUMAN ECOLOGY	15
	Project 1-1: How is your environment important to you?	15
	Project 1-2: Analyzing the use of critical land in your community as building sites.	17
	Project 1-3: How to estimate the amount of runoff from developed land.	23
	Project 1-4: Determining the efficiency of parking facilities.	25
	Project 1-5: Determining how the average speed of cars is related to the volume of traffic.	28
	Project 1-6: Determining human population density.	30
	Project 1-7: How much electrical energy is used in your home?	31
	Project 1-8: Is there adequate open space in your community?	33
	Human Ecology Questions	35
2.	ENVIRONMENTAL FACTORS	37
	Project 2-1: How light affects plant growth.	37
	Project 2-2: How etiolation affects the ratio of dry mass to gross mass.	39
	Project 2-3: How variation in light intensity helps determine the kinds of plants that can grow in an area.	40
	Project 2-4: Effect of temperature changes on the breathing rate of goldfish.	44
	Project 2-5: The effect of temperature on the metamorphosis of mealworm pupae.	45
	Project 2-6: How temperature affects sprouting of seeds.	46
	Project 2-7: Temperature difference between a north-facing slope and a south-facing slope.	48
	Project 2-8: How do variations in annual rainfall affect the rate of growth of trees?	50
	Environmental Factors Questions	52
3.	NATURAL CYCLES AND TRANSFER OF ENERGY	53
	Project 3-1: How the leaves of a plant affect the rate of transpiration.	53
	Project 3-2: How relative humidity affects the rate of transpiration.	55
	Project 3-3: Why it is cool under a tree.	56
	Project 3-4: How an animal may affect the carbon dioxide-oxygen cycle.	57

Project 3-5: How oxygen is produced during photosynthesis.   58
Project 3-6: How is an animal's increase in mass related to the mass of the food it eats?   60
Project 3-7: What is the percentage of minerals in leaves?   62
Project 3-8: How much mineral matter does soil contain?   64
Natural Cycles and the Transfer of Energy Questions   66

4. NICHES, TERRITORIALITY, AND THE BALANCE OF NATURE   67
Project 4-1: Some food preferences of insects.   67
Project 4-2: Food preferences of the cabbage butterfly caterpillar.   69
Project 4-3: Checking on specialization in gall insects.   71
Project 4-4: The niche of the bee.   72
Project 4-5: Succession in a molasses solution.   74
Project 4-6: Paramecium and Didinium—Balance of Nature in action.   76
Project 4-7: How the robin claims and defends its territory.   78
Project 4-8: Territoriality among yellow jacket hornets.   79
Niches, Territoriality, and the Balance of Nature Questions   84

5. COMMUNITIES   85
Project 5-1: Which species make up the community in a rotten log?   85
Project 5-2: How to determine similarity and dissimilarity among plants in different locations.   88
Project 5-3: Judging the diversity of plants in a field.   89
Project 5-4: Distribution of the spittlebug.   91
Project 5-5: Estimating the population of spittlebugs in a field.   92
Project 5-6: The capture-release-recapture method of estimating population.   93
Project 5-7: The reproductive potential of goldenrod.   95
Project 5-8: Squirrels, acorns, and oaks.   96
Communities Questions   99

APPENDIX   103
Project A-1: How to obtain topographic maps.   103
Project A-2: Making a plane table survey.   103
Project A-3: How to lay out a transect.   109
Project A-4: Making a quadrat frame.   109
Project A-5: How to measure relative humidity.   112
Project A-6: Differentiating plants.   113

ECOLOGY TEXTBOOKS   135

INDEX   137

INTRODUCTION

Look into your environment. It consists of everything that affects you. Some effects of your environment are beneficial and give you pleasure. Some others are annoying or even harmful.

What gives you pleasure?

What annoys or irritates you?

Is the air you breathe pure and health-giving? Or is it polluted and poisoned by automobile exhaust, smoke, and chemical fumes?

Do you have easy access to nearby recreation areas? Or must your sports and recreation take place in far-off areas or on unsafe streets?

Are the streams, lakes, or other bodies of water in your area pure enough for swimming and sport fishing?

These are only a few of the many questions that are concerned with your relationship with your environment. They are the kinds of questions you would investigate in ecology, for ecology is the science of the way organisms (and people are organisms) relate to their environment. By investigating ecology you may learn the causes of annoying problems and what you can do about them.

A wise person once said "Science is a verb." Science is active, not passive. You learn science by carrying out scientific projects and experiments. You must perform the procedures accurately and observe carefully. Reading about science, however, is also important since it gives you some of the background information you need. Reading about science may also stimulate your imagination so that you become aware of new problems and are led to design procedures for solving them.

Investigating Ecology is based on these principles. All its projects are concerned with ecological problems. To prepare you for performing each project, you are given some background information. Specific instructions for carrying out the project are then presented. In the section **For Further Investigation** there are suggestions for other projects which might interest you.

Investigating Ecology is *open-ended*. The solution of each problem reveals other problems. Only a few of these are listed in **For Further Investigation**; there is almost no limit to the number of additional related problems you might think of. When a new problem occurs to you, write it down. You may not immediately have the time to design and perform the experiment required to solve the problem, but perhaps at some later date you will.

Becoming aware of a problem is possibly the most important part of science. Designing and performing the experiment are much more routine. Yet the successful scientist is the person who actually designs and carries experiments to their conclusions after he has realized that the problems exist. Scientific problems occur to many people; it is the scientist who tries to solve them.

Investigating Ecology will have achieved its goal if it helps make you more aware of your environment, stimulates you to solve some of the ecological problems you perceive, and provides you with some techniques for their solution.

INVESTIGATING ECOLOGY

Fig. 1-1. Many things in your environment can give you pleasure. (Courtesy American Airlines)

Chapter 1

HUMAN ECOLOGY

What is there in your surroundings that gives you pleasure? What irritates and annoys you? What may endanger your well-being? Your greatest pleasures and your most irritating annoyances are often the results of how you react to your environment. When once we become aware of these influences on our well-being, we can take the first steps to make our daily lives more comfortable and satisfying. To do this we must first examine those factors around us to which we react. When we understand the nature of these factors, we can then proceed to relate to our environment in the most effective and satisfying ways.

How does your own environment affect you personally?

Project 1-1 lists some of the ways in which the environment affects people. Which of these means the most to *you*? Only you can answer this question. People differ in their tastes and preferences. What one person considers extremely important may not interest another at all. By completing Project 1-1 you can discover and record your own feelings about the way your environment affects you personally.

PROJECT 1-1: How is your environment important to you?

You will need: a notebook and a pen.

1. In your notebook, copy the numbers that are printed before the environmental effects.

 1. Sport fishing
 2. Commercial fishing
 3. Hunting
 4. Trapping
 5. Swimming
 6. Pleasure boating
 7. Water transportation
 8. Skiing
 9. Ice skating
 10. Ball playing
 11. Hiking
 12. Rock climbing
 13. Bird watching
 14. Nature study
 15. A large variety of plants
 16. A large variety of animals
 17. Beautiful flowers
 18. Scenic beauty
 19. Control over populations of harmful animals (insects, rodents, etc.)
 20. Plentiful supply of food for wild animals (birds, mammals, fish, etc.)
 21. Suitable habitats for wild animals
 22. Plentiful supply of nutrients for plants
 23. Plentiful supply of oxygen for living organisms
 24. Plentiful supply of fresh water for irrigation
 25. Plentiful supply of unpolluted drinking water
 26. Adequate rate of recharge of aquifers (underground water supplies)
 27. Prevention of flooding
 28. Locations to receive effluents from sewers
 29. Prevention of soil erosion
 30. Ores and other minerals

31. Fossil fuels
32. Unpolluted air
33. Plentiful supply of energy
34. Building sites
35. Uncrowded communities
36. Superhighways
37. Quiet
38. Plentiful supply of food for people
39. Plentiful supply of lumber
40. Fast, convenient transportation
41. Prevention of disease
42. Comfortable indoor temperatures
43. Shade trees
44. Suitable level of human population
45. Adequate employment opportunities

2. Alongside each number write the letter that indicates your degree of interest in that specific environmental effect.

a—great interest   **c**—little interest
b—considerable interest   **d**—no interest

3. To compare your list of interests with those of other members of your class, offer to write your list on the chalkboard. Then other students can write their ratings alongside. In this way you can get a general idea of what most students think is important and how the students' opinions differ.

4. Discuss the reasons for the differences of opinion with your classmates.

For Further Investigation

1. Make a list of some other schools in which this book is used. Correspond with the classes in these schools in various types of communities (inner city, urban, suburban, rural) that have completed Project 1-1. Compare the results of the project in the different schools.

2. Determine whether the girls' responses in your class differ from those of the boys and, if they do, what the differences are.

3. Add more items to the list of things you would like your environment to provide.

4. How might the results differ if the project were completed by an automobile manufacturer? A farmer? A vacation resort owner?

In Project 1-1 you recorded the extent of your interest in a number of environmental effects. The other projects in this book will help you discover how environmental factors affect organisms, including humans. Your increased knowledge about your environment will enable you to decide what you can or should do to help improve your environment.

People disrupt the environment when they convert forests and prairies into agricultural land. Areas that previously were suitable for a wide variety of organisms have been so changed that the plants and animals that formerly lived in the area can no longer survive there. The destruction of forests and tough, natural prairie grasses has caused soil erosion, floods, and siltation of rivers and lakes. It has decreased our water supply and spoiled our environment in many other ways. Yet people continue to create serious problems for themselves as well as for other forms of life by destroying the natural conditions that protect all living things against catastrophe.

We cannot, however, completely avoid disrupting the environment. If humans are to exist in at least their present numbers, they must have farms in place of some forests and prairies. When it comes to farmland, we cannot always choose the ideal location but must adapt to existing conditions. In some cases, however, we do have a rather wide choice of the kinds of habitats we can change and develop for other purposes. This is especially true in choosing land for building construction. It is not always necessary to build in a *particular* location. Any one of a number of other places might be quite as suitable. Still, we frequently build on what is called "critical land." These are areas where buildings should not be constructed. Severe environmental damage is caused by the drastic changes accompanying real estate development of such land. Critical areas include steep slopes on hills or mountains, swamps or marshes, floodplains onto which rivers may overflow, and locations where there is only a thin layer of soil (or none at all) over the bedrock. Even if the environmental damage is ignored, critical areas cannot be considered suitable sites for buildings. Homes built on filled-in swamps, for example, will almost inevitably have wet cellars.

Are there any structures in your own community built on critical land? We will find out in Project 1-2.

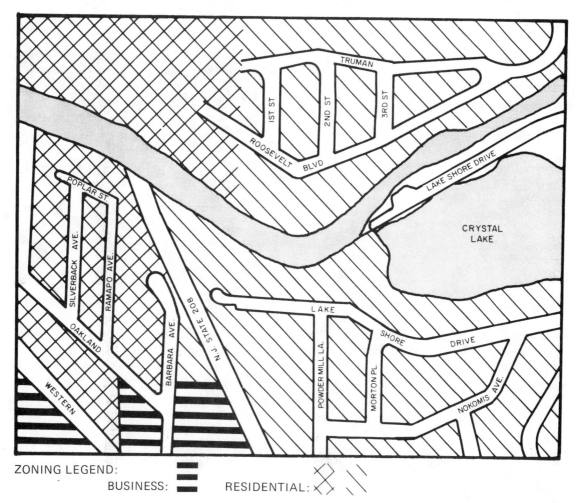

Fig. 1-2. A section of a real estate map of Oakland, N.J.   *(Courtesy of Borough of Oakland)*

PROJECT 1-2: Analyzing the use of critical land in your community as building sites.

You will need: a topographic map of your area (see Project A-1 in the Appendix: How to obtain topographic maps), a real estate development (or zoning) map of your area (which may be obtained from the real estate tax assessor's office), a floodplain map (which may be obtained from U.S. Geological Survey, Water Resources Division, NE Region, Reston, Virginia 22092), a camera, color slide film for the camera, a slide projector, a magnetic compass, a ruler, a set of transparent markers in yellow, red, and green, a strip of cardboard, a notebook, a pencil, and a pen.

1. Mount the real estate development map on a wall (Fig. 1-2).

2. Locate the portion of the topographic map that covers the area you wish to analyze. Photo-

graph that part of the map, using color slide film (Fig. 1-3).

3. After the slide has been developed, project it onto the real estate development map. Move the projector toward or away from the map until identifiable features (streams, ponds, roads, buildings, etc.) on the real estate development map and the projected topographic map coincide.

Fig. 1-4. Tracing contour lines onto real estate map.
(Photo by Alan J. Federow)

map to determine the distance that corresponds to 100 feet. Lay out this distance several times along the edge of a cardboard strip. By placing this measuring device on the map perpendicularly across some contour lines, you can determine whether the contour lines are closer together than the lines on your measuring device (making the slope greater than 10%), farther apart (making the slope less than 10%), or spaced exactly the same (in which case the slope is equal to 10%).

Determine which parts of the area have a slope of 10% or more. Color these parts with a yellow marker.

6. Color the swampy or marshy areas green.

7. Photograph the floodplain map of your area, using color slide film.

8. After the slide has been developed, project it onto the real estate development map. Adjust the projector, as in Step 2, so that identifiable features on the floodplain map and the real estate map coincide.

Fig. 1-3. Photographing topographic map.
(Photo by Alan J. Federow)

4. Trace contour lines, streams, ponds, and swamps from the topographic map onto the real estate development map, unless those features are already shown (Fig. 1-4).

5. A slope is 10% when 10-feet contour lines are 100 feet apart. Ten feet (the vertical rise from one contour line to the next higher one on some topographic maps) is 10% of 100 feet (the horizontal distance between the two contour lines). You can make a measuring device (Fig. 1-5) on which lines are spaced at a distance that represents 100 feet on the map. Refer to the scale on your

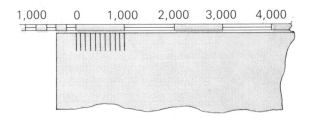

Fig. 1-5. Slope-measuring device on unlined file card to be used with topographic map. Lines spaced at distances of 100 feet.

Fig. 1-6. Floodplain map of same part of Oakland as in Fig. 1-2. Oakland is located in a hilly area. The borders of the red-tinted area indicate the highest flood line within 100 years. Houses located in this red-tinted area are likely to suffer from flood damage during prolonged rainfall. (*Courtesy of Borough of Oakland*)

9. Trace the outlines of floodplains onto the real estate development map. Color the floodplains red (Fig. 1-6).

10. After you have colored each type of critical area, count and record the number of buildings in each—excessively sloping land, swampy or marshy land, and floodplain. Note that these critical areas may overlap to some extent.

11. Count and record the total number of buildings on the real estate development map of the area you are analyzing.

12. Count and record the total number of buildings constructed on critical areas. Because of overlapping of different types of critical areas, this total may not equal the sum of the numbers of buildings on each type of critical area. Buildings in one kind of critical area may also appear in some other classes of critical areas. (See Step 9.)

13. Calculate the percentage of buildings in the analyzed area which are constructed on
   a. excessively sloping land
   b. swampy or marshy land
   c. floodplains
   d. critical land of any type.

14. In the area you have investigated, has the land development been for the best interests and welfare of all?

For Further Investigation

1. Conduct a survey similar to that in Project 1-2 but on the basis of the area of critical land that has been developed as sites for buildings rather than the number of buildings constructed.

2. Find out the amount of undeveloped critical land in your community and the kinds of uses for which the land is zoned. A real estate zoning map of the area will indicate the permitted uses for every part of the area.

3. Develop an appropriate questionnaire and conduct a survey of homeowners to determine if houses that are built on the critical land you investigated in Project 1-2 have such problems as wet cellars, flooding, sewage difficulties, and erosions (Fig. 1-7).

*Fig. 1-7. Conducting a survey of homeowners.
(Photo by Alan J. Federow)*

4. Find out what kinds of laws (national, state, and local) have been passed to prevent building construction on critical land.

5. Correspond with students who have conducted Project 1-2 in other localities. Compare results.

Think About This Problem

A few years ago a number of homes in California were completely destroyed when they slid down steep hillside slopes.

How might this problem affect you?
What caused the problem?
How can you help solve it?

Developing critical land, by constructing buildings, roads, or other facilities, can harm the environment in many ways. Trees and other plants must be removed to permit land development. This destroys the complicated system of roots that holds soil particles together and eliminates the leafy natural umbrella that protects the

*Fig. 1-8. Soil erosion on recently constructed road.
(Courtesy American Museum of Natural History)*

ground from the full force of falling rain. As a result serious soil erosion frequently occurs (Fig. 1-8). Draining swamps, in preparation for development, lowers the water table (the level below which soil is completely saturated with water). See Fig. 1-9. This wastes water that would otherwise enter wells. The wells supply water for private residences as well as for entire towns. Habitats suitable for ducks, deer, muskrat, crayfish, and many other species are destroyed when swamps are drained.

Building construction on a floodplain is generally followed by urgent demands by homeowners that something be done to prevent flooding, even though flooding is natural and is to be expected on floodplains (Fig. 1-10). In response to the demands, the authorities may "channelize" or straighten the stream, build levees, and deepen the stream by dredging. What remains is actually a ditch that has none of the beauty of the stream and is unsuitable as a habitat for wildlife. Despite such work and the destruction of the natural qualities of the stream, flooding still occurs and the original problem remains.

It would be impossible to describe all the harm done by development of critical land because the damage is so varied. Only land that is not in the critical category should be developed;

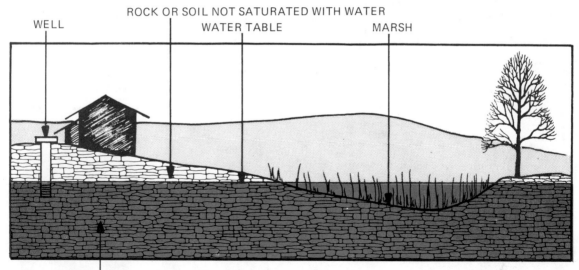

Fig. 1-9. *Cross section of a water table.*

Fig. 1-10. *This homeowner never expected a flood even though his home is located on a floodplain.*
(Courtesy U.P.I.)

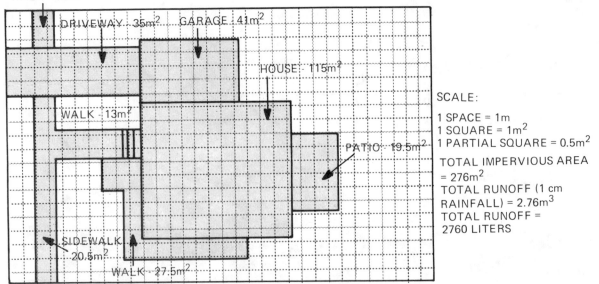

Fig. 1-11. Plan of a building on lot, showing how to calculate volume of water that runs off impervious surfaces (shaded) during a 1-centimeter rainfall.

Fig. 1-12. Property damage and personal danger can result from building on floodplains.
(Courtesy World Wide Photos)

yet even on noncritical land, development can result in many kinds of environmental damage. For example, land development may help cause severe floods. When land is developed, some of its area is generally paved and houses are usually constructed. These structures are impervious and prevent rainwater from entering the ground. Instead, the rainwater becomes runoff which flows along the impervious surfaces (Fig. 1-11), rushes into storm sewers, and finally into rivers or other bodies of water where it greatly increases the chance of flooding (Fig. 1-12). In Project 1-3, you will investigate the way development affects runoff during a rainstorm.

PROJECT 1-3: How to estimate the amount of runoff from developed land.

You will need: equipment for a plane table survey (see Project A-2 in the Appendix: Making a plane table survey), a notebook, and a pen.

1. Select an area where you wish to estimate the amount of runoff that may be expected during a rainstorm. The area may include buildings, roads, sidewalks, driveways, parking areas, etc.
2. Make a map of the area on ordinary graph paper, using the plane table method (Fig. 1-13) described in Project A-2 in the Appendix. Let each square of the graph paper represent a convenient length, say ½ meter (Fig. 1-14).
3. Determine and record the area covered by paving, buildings, or other impervious structures. This can be done by counting squares on the graph paper. Count as ½ square any square that is partially inside the outline of the impervious area.
4. Calculate the number of cubic meters of water that fall on the impervious area during a 1-centimeter rainfall.
5. Calculate the number of liters of runoff water represented by the result you obtained in Step 4. One cubic meter equals almost exactly 1000 liters.

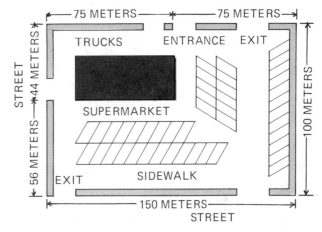

Fig. 1-14. Plan of a typical parking area.

For Further Investigation

1. Find out the number of kilometers of paved streets and roads in your community, as well as their average width. Your highway department can give you this information. Calculate the number of liters of water that fall on the streets and roads during a 1-centimeter rainstorm (Fig. 1-15).

2. To find out what your community does with runoff water, investigate the following questions:
   a. How are storm sewers different from sanitary sewers?
   b. Why shouldn't storm runoff water be permitted to flow into the sanitary sewer system?
   c. What are the functions of catch basins?
   d. What are the diameters of the pipes used in the storm sewer system (if your community has one)? How are the sizes of the pipes related to the rates at which water flows through them?
   e. What maintenance procedures are used to keep the storm sewers in proper operating condition?
   f. Where does the runoff water go when it flows from the storm sewer outfall? How might the runoff water increase the possibility of flooding?

Fig. 1-13. Plane table survey of parking lot.
(Photo by Alan J. Federow)

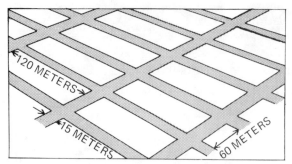

Fig. 1-15. Calculating the liters of water falling on impervious street surface during a 1-centimeter rainstorm. Determine the length and width of streets as shown here. Measure only the shaded areas.

3. Flooding occurs when water enters a stream more rapidly than it can flow through the stream. The excess water flows across the adjacent land. Determine the rate, in cubic meters per minute, at which water flows in a brook or other small stream. A rough estimate may be made in the following way: Make a drawing of a cross section of the stream (Fig. 1-16). Calculate the cross-sectional area of the stream. Use a chip of wood, a tape measure, and stopwatch to determine the velocity of the water in the stream. Multiply the cross-sectional area of the stream by the distance the water flows in one minute in order to obtain the volume of flow per minute.

    a. From a bridge or plank laid across a stream measure the width of the stream from one edge to the other.

    b. Indicate this width on graph paper, allowing each space on the graph paper to represent $1/10$ meter.

    c. With the meter stick measure the depth of the stream at each successive $1/10$ meter along the width of the stream and mark these measurements on graph paper.

    d. Join these marks together to form a drawing of the cross section of the stream.

    e. To determine the area of the cross section, count the number of squares. (Each square represents $1/100$ square meter.) Count each partial square as $1/200$ square meter. Add the complete squares and the partial squares to obtain the total number of square meters in the area of the cross section.

4. Runoff water from some storm sewer systems is permitted to flow into holding ponds that help

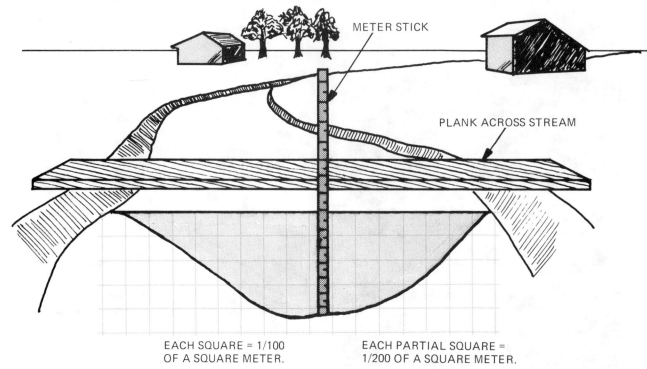

Fig. 1-16. Drawing showing how to calculate the cross-sectional area of a stream. The total cross-sectional area of this stream is 0.355 square meters.

prevent flooding of streams. Calculate the volume of runoff water that would raise the water level in such a pond by 1 meter if the pond is 1 kilometer square.

Think About This Problem

A stream that rarely flooded before the construction of a large real estate development on its watershed floods frequently now that the development has been completed.

How might this problem affect you?
What caused the problem?
How could you help solve it?

Paving a piece of land or building a house on it prevents water from seeping into the soil. Instead, the water becomes runoff that generally goes into storm sewers. The development of a large area can cause serious flooding when the runoff enters streams during a storm. If the pipes used in a storm sewer are too narrow, they cannot drain off runoff water rapidly enough. In that case, all the sewer pipes in the line remain full of water, even though water is flowing through them as fast as possible. No additional runoff water can enter the sewer openings and, in fact, back-up water from uphill locations may spout from the openings. The results are flooded roads, flooded cellars, soil erosion, and a host of other troubles.

In communities that have no storm sewer systems, storm water simply runs downhill, often flooding cellars and creating temporary ponds on lawns and fields in lower areas. Roadside ditches handle some of the flow but many of them are inadequate. Water flowing rapidly through ditches may undermine roads, making them dangerous to ride on and increasing maintenance costs.

Much of the trouble may be avoided by planning construction in such a way that a maximum of rainwater is permitted to enter the soil. Rainwater that falls on roofs should flow from the gut-

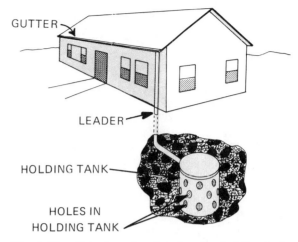

Fig. 1-17. Cut-away drawing showing holding tank below the surface of the ground.

ters and leaders into dry wells (porous, underground holding tanks) which permit the water to soak into the soil slowly (Fig. 1-17). Similarly, water from paved areas, such as driveways and parking lots, should be directed into dry wells. Porous paving material should be used where possible. Storm sewer outfall pipes should lead water into holding ponds or swamps, from which a large percentage of water can seep into the ground instead of flowing into streams which may consequently flood.

Soil has an enormous capacity for holding water and thus decreasing the probability of flooding. In fact, some soils can hold more than their own weight in water. The soil covers most of the area in any community and under the right conditions, additional water can temporarily saturate the soil from the surface to the water table. That is why the soil can accept truly huge quantities of storm water.

A large part of the storm runoff problem in some communities is created by parking lots. One way of reducing this adverse effect of automobile parking is to use more efficient methods of parking cars. You will investigate the efficiency of various kinds of parking facilities in Project 1-4.

PROJECT 1-4: Determining the efficiency of parking facilities.

You will need: a metric tape measure, chalk, a notebook, and a pen.

1. Locate at least one of each of the following kinds of parking facilities: a garage belonging to a one-family or two-family house, a multistory parking garage in which attendants park the cars, a

multistory parking garage where cars are parked by their owners, a parking lot where attendants park their cars, a parking lot where cars are parked by their owners, a stretch of curbside parking. Record the locations of these parking facilities for future reference.

2. By making measurements (Fig. 1-18), or by interviewing the managers, determine and record the ground surface area in square meters covered by each of the parking facilities. Include the area of the driveway, if any.

3. Find out and record how many cars can be parked at each parking facility. You may be able to do this in some cases by counting parking spaces; in other cases you will have to interview the manager of the parking facility. Find out and record the number of floors in each multistory parking garage.

4. For each parking facility, calculate the efficiency in terms of the number of cars that can be parked per 100 square meters of ground surface area. To compare the efficiencies of attendant-parking multistory garages and owner-parking mutistory garages that do not have the same number of floors, calculate what the efficiencies of the garages would be if both had the same number of floors. For example, you may have calculated the efficiencies of a four-story attendant-parking garage and a three-story owner-parking garage. Multiply the efficiency of the owner-parking garage by 4 and divide the product by 3.

5. Construct a table showing a comparison of the efficiencies of the different kinds of parking facilities you analyzed. Arrange the data as shown in the sample table in Fig. 1-19.

6. Explain the environmental benefits and disadvantages of high-efficiency parking facilities.

For Further Investigation

1. Determine whether or not a parking lot owner makes a greater profit by employing parking attendants. Do so by estimating income and expenses for the two kinds of parking lots.

Fig. 1-18. Measuring an attended parking lot.

(Photo by Alan J. Federow)

Comparative Efficiencies of Different Kinds of Parking Facilities

TYPE OF FACILITY	ACTUAL EFFICIENCY	* ADJUSTED EFFICIENCY
Residential garage		
Multistory garage, attendant parking		
Multistory garage, owner-parking		
Lot, attendant parking		
Curbside parking		

* Adjusted efficiency presents the efficiencies of multistory parking garages as if all had the same number of stories as the tallest garage.

Fig. 1-19. Sample table shows one way to arrange data on efficiencies of different kinds of parking lots.

2. Some communities provide automobile owners with facilities in outlying areas where they can park their cars, then use free mass transportation to travel to the business district. Investigate the environmental benefits and disadvantages of such a system.

Think About This Problem

Traffic has generally been moving very slowly on a certain street in a downtown business district because of cars constantly parked along both curbs.

How might this problem affect you?
What causes the problem?
How could you help solve it?

Parking undoubtedly causes environmental damage but it is actually a lesser problem when compared to other ways automobiles harm the environment. Automobile exhaust contains a variety of pollutants which directly harm humans and other organisms; the pollutants also damage abiotic (nonliving) environmental factors, such as air, water, soil, and light. Exhaust adds to the air such poisonous substances as sulfur oxides, nitrogen oxides, unburned hydrocarbons, and carbon monoxide. With the aid of sunlight, nitrogen oxides and hydrocarbons convert oxygen to ozone, which is a poisonous gas. Peracyl nitrate and aldehydes are other dangerous chemicals formed in this same set of reactions.

Nitrogen oxides in the air become nitric acid, which forms nitrates after it enters either the soil or bodies of water (Fig. 1-20). Nitrates improve the fertility of the soil; however, they stimulate a rapid growth of algae when the nitrates are added to bodies of water, thus spoiling the water for swimming or fishing. These are but a few of

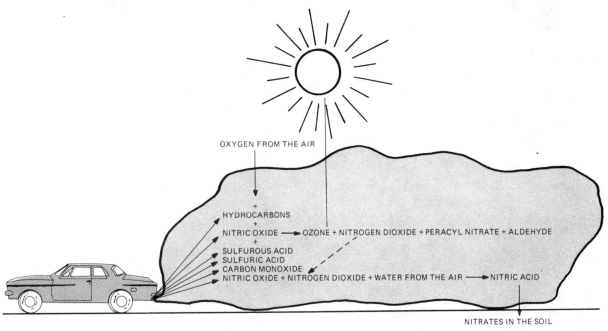

Fig. 1-20. How solar energy helps automobile exhaust combine with oxygen and water to pollute the atmosphere.

the detrimental effects of automobile exhaust on abiotic environmental factors.

The harm that exhaust does to humans is very serious indeed. Diseases traceable to automobile exhaust include lung cancer, emphysema, bronchitis, anemia, heart disease, and lead poisoning.

A concerted effort is being made to decrease the amount of dangerous exhaust components each automobile emits per kilometer. Catalytic converters, better ignition systems, and other devices have been successful in decreasing harmful exhaust emissions. Even with the greatest improvement that can be expected, however, automobile exhaust will still do damage. For this reason, it is important that other methods also be used to help decrease emissions. One practical method is to reduce the number of cars on our roads and streets, especially where traffic is particularly heavy now. As traffic volume is reduced in an effort to decrease pollution, there will probably be an increase in the average speed of the cars. In Project 1-5 you will determine how average speed is related to volume of traffic.

PROJECT 1-5: Determining how the average speed of cars is related to the volume of traffic.

You will need: a stopwatch, a metric tape measure, a pair of walkie-talkies, a notebook, and a pen.

1. Select a location on a street where traffic is heavy (Point I).
2. Select a location on a street where traffic is relatively light but where cars frequently pass (Point II).
3. Count and record the number of cars that pass Point I in 10 minutes (Fig. 1-21). Calculate the rate of traffic flow in terms of cars per minute.
4. As soon as possible after Step 3 has been completed, repeat Step 3 for Point II.
5. One person (designated as A) is to stay, with a stopwatch and walkie-talkie, at the place where the survey of heavy traffic was made at Point I. Measure a distance ¼ kilometer down the street from the starting point. Another person (designated B) is to be stationed at this distance with a walkie-talkie.
6. A is to select a car that is about to pass the starting point. As it does so, he is to start the stopwatch, then describe the car (make, color, license plate number, other distinctive features) to B.
7. When the car passes B, he is to say "Stop" to A and then describe the car in order to prevent mistakes (Fig. 1-22).
8. A is to record the time for the ¼-kilometer run (Fig. 1-23).
9. Repeat Steps 6, 7, and 8 several times at the location where the heavy volume of traffic was surveyed.
10. Repeat Steps 5, 6, 7, 8, and 9 for the location where the light traffic was surveyed.
11. Calculate the velocity in kilometers per hour for heavy traffic and for light traffic.

NOTE: Velocity in kilometers per hour may be calculated by using the formula:

$$\text{Velocity} = \left(\frac{900}{x}\right)\left(\frac{\text{kilometers}}{\text{hour}}\right)$$

where time for ¼ kilometer = x seconds.

The formula is derived as follows:

$$\text{Velocity} = \frac{\text{distance}}{\text{time}}$$

$$\text{Velocity} = \frac{\text{¼ kilometer}}{x \text{ seconds}}$$

$$\text{Velocity} = \frac{\text{kilometer}}{4x \text{ seconds}}$$

Fig. 1-21. Counting and recording the number of cars that pass point I.

Fig. 1-22. Car passes B.

$$\text{Velocity} = \left(\frac{\text{kilometer}}{4x \text{ seconds}}\right)\left(\frac{3600 \text{ seconds}}{\text{hour}}\right)$$

$$\text{Velocity} = \left(\frac{900}{x}\right)\left(\frac{\text{kilometers}}{\text{hour}}\right)$$

Example: It takes 45 seconds to travel ¼ kilometer.
What is the velocity in kilometers per hour?

Fig. 1-23. A records the time.
(*Photos by Alan J. Federow*)

$$\text{Velocity} = \left(\frac{900}{x}\right)\left(\frac{\text{kilometers}}{\text{hour}}\right)$$

$$\text{Velocity} = \frac{900}{45} \frac{\text{kilometers}}{\text{hour}}$$

$$\text{Velocity} = \frac{20 \text{ kilometers}}{\text{hour}}$$

12. Explain your findings.

For Further Investigation

1. Do reading research to learn the quantities of harmful exhaust components that the average automobile emits for each liter of gasoline that it burns. Use this information to calculate the rate at which these components enter the air at each of the two places where you analyzed the traffic flow in Project 1-5.

2. Repeat Project 1-5 at the same locations at different times of day and different days of the week. Graph your findings.

3. Conduct traffic surveys to find out how the volume of traffic is influenced by such factors as highway entrances and exits, bridges, shopping centers, and the number of lanes in the roadway.

4. Count and record the number of passengers in each car that passes a certain point. Calculate the average number of passengers per car. What conclusions can you draw about reducing the volume of traffic?

Think About This Problem

Traffic policemen stationed at a busy intersection have severe anemia and show signs of emphysema even though they do not smoke excessively.

How might this problem affect you?
What causes the problem?
How could you help solve it?

Many factors influence the volume of traffic. People drive mainly to get to and from work, shopping areas, entertainment centers, or to visit family and friends. The use of buses, trains, and other forms of mass transportation can reduce the need for so many cars on the roads. Car pooling is another effective method of reducing traffic volume. A strategy that has recently been receiving greater attention is the "back to the city" movement. It is hoped that many people who now live in the suburbs and commute to and from a city will decide to live in the city if suitable homes are made available. The volume of traffic would be dramatically reduced if almost everyone who is employed in a city or uses its facilities lived there.

Planners must know whether it is possible to provide enough residences, either by renovation or new construction, for the large numbers of people they hope to attract to a city. Knowing the population density for various kinds of residential developments permits the planners to approach the problem intelligently. The methods they use to determine population density are somewhat similar to those in Project 1-6, in which you will calculate the population density for an area of your community.

PROJECT 1-6: Determining human population density.

You will need: a 20-meter tape measure, real estate development maps of your community, a notebook, and a pen.

1. Select a part of your community in which you wish to determine the human population density.
2. Measure the dimensions, in meters, of the selected part of your community. Measure from the middle of each road or street that borders the selected piece of ground (Fig. 1-24).
3. Calculate the area of the piece of ground in square meters. If it is rectangular, the area can be determined by multiplying the length by the width. It is generally necessary to refer to real estate development maps to determine the areas of pieces of ground that are not rectangular. Convert the area from square meters to square kilometers by dividing by 1,000,000.
4. Find out and record the number of people living in each building in the area you have selected. This may be done by interviewing the owners or managers of the buildings or by referring to local government records.
5. Calculate the total number of people living in the selected area.
6. Divide the total number of people by the area in square kilometers to obtain the population density in people per square kilometer.

For Further Investigation

1. Conduct projects similar to Project 1-6 in neighborhoods where there are the following types of residences:
    a. high-rise apartment houses (12 stories or higher)
    b. apartment houses less than 12 stories tall
    c. tenement buildings
    d. attached town houses
    e. detached two-family houses
    f. detached one-family houses

If your community does not include each of

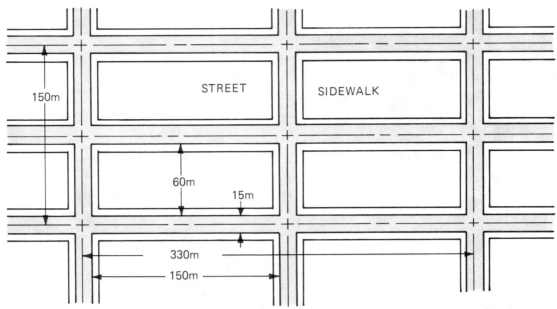

Fig. 1-24. Layout of streets in a typical community for the purpose of measuring population density.

the types of residences listed above, correspond with students in schools in communities that do have such buildings and suggest that they conduct the surveys and forward the data to you.

Compare the population densities for each type of residence. Explain your findings.

2. Find out how many people commute each day to and from the urban center in or near your community. Your local government may have this information. Calculate the number of square kilometers of land that would provide living space for all the commuters if they lived in residential buildings like those in the urban center.

3. Do reading research to learn the population densities of various cities, states, and countries.

4. Describe the advantages and disadvantages of high human population density in the urban centers of countries that are not overpopulated as a whole.

Think About This Problem

Land costs $40,000 per acre in a certain community where the zoning laws prohibit any kind of residence except one-family, completely detached houses and require that each house be built on no less than 1 acre of land.

How might this problem affect you?
What caused the problem?
How could you help solve it?

Many city dwellers hope to move to the suburbs. Some who achieve this ambition, however, soon move back again to the city. Evidently, they miss something that the city provides. The high human population density that is characteristic of cities offers numbers of advantages. For many people, these advantages compensate for the disadvantages of city life. One readily measurable advantage is that less energy is needed per person in cities than in suburban areas. You will investigate one aspect of the energy requirement in Project 1-7.

PROJECT 1-7: How much electrical energy is used in your home?

You will need: a notebook and a pen.

1. List every electrical device that works on house current (supplied by the electric utility company) in your home. Record the wattage of each electrical device and the approximate average

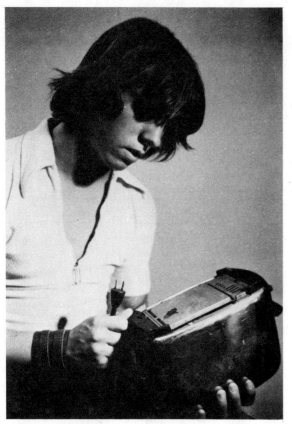

Fig. 1-25. Reading wattage at the bottom of electric toaster. Be sure that plug is not in outlet.
(Photo by Alan J. Federow)

Fig. 1-26. Reading electric bills.
(Photo by Alan J. Federow)

number of hours per day it is used (Fig 1-25).

2. Calculate the number of kilowatt-hours (KWH) of electrical energy each device consumes per month, then add the individual numbers of KWH to estimate the total amount of KWH used per month in your home. The number of kilowatt-hours a device uses per month is equal to its wattage, multiplied by the number of hours it is used per day, multiplied by 30, and divided by 1000. The formula is:

$$\text{KWH} = \frac{\text{Watts} \times \text{hours per day} \times 30 \text{ days per month}}{1000}$$

4. Read some of your recent electricity bills to learn the actual number of KWH of electricity used per month in your home (Fig. 1-26).

5. Explain any difference between the number of KWH you obtained by estimation and the actual number of KWH as shown on your bills. Is the method of estimation described in this project accurate enough to be of practical use?

3. Construct a table similar to the following:

DEVICE OR FIXTURE	WATTAGE (A)	HOURS PER DAY (B)	KWH PER MONTH $\frac{A \times B \times 30}{1000}$
Lamp Bulb	100 watts	6	18
TOTAL	XXXXX	XXXXX	18

For Further Investigation

1. Find out the total number of KWH sold in each of the last several years by the company that supplies electricity to your community. Make a graph of number of KWH per year versus time.

2. Find out the amount of heating fuel used per resident per year in at least one of each of the following types of residences:
   a. multistory apartment building
   b. attached townhouses
   c. detached two-family houses
   d. detached one-family houses.

   Compare your findings for each type of residence. What evidence have you found to support or rebut the idea that less energy is required per person in communities where human population density is higher?

3. List the ways in which a person decreases the amount of energy he uses after moving from the suburbs to the city in which he works and whose facilities he uses.

Think About This Problem

Certain communities have experienced "brownouts" or "blackouts" because their supply of electrical power was too small to satisfy the demand at certain times of the year.

How might this problem affect you?
What caused the problem?
How could you help solve it?

People who live in cities would like to have at least some of the advantages of the suburbs, even if they would not actually want to move to the suburbs. One condition most people appreciate and want is open space, whether it is in the form of playgrounds, tree-filled parks, or undeveloped land. Because urban land is so expensive, there is usually very little open space unless some truly foresighted people insisted that it be set aside as the city grew. Your community may be fortunate enough to have ample open space. Or on the other hand, there may be far too little open space to meet the needs of the people in your community. In Project 1-8 you will examine the open space situation in your community.

PROJECT 1-8: Is there adequate open space in your community?

You will need: a topographic map that includes your community (see Project A-1 in the Appendix: How to obtain topographic maps), a metric ruler, a set of colored crayons, a notebook, a pencil, and a pen.

1. On the topographic map locate all the open spaces in your community.
2. Around each park or other open space, draw lines parallel to the boundaries of the open space and at a distance that represents ¼ kilometer. Refer to the scale on the map to determine the length representing ¼ kilometer. Use a red crayon to fill in the space between the lines you drew and the boundaries of the open spaces.
3. Repeat Step 2 for distances of ½ kilometer, 1 kilometer, 2 kilometers, and 3 kilometers. Use a different colored crayon for each distance. Stop drawing a line around one open space when it touches a line drawn around any other open space.
4. Determine and record the percentages of the community's total area that are within ¼ kilometer of an open space, ½ kilometer of an open space, etc.
5. Estimate the time required to walk ¼ kilometer, ½ kilometer, 1 kilometer, 2 kilometers, and 3 kilometers.
6. Do you believe your community has enough parks and other spaces in the right locations? Give your reasons for your answer. If there is adequate open space, correctly distributed, you should be

able to walk to an open space in a reasonable length of time (Fig. 1-27).

For Further Investigation

1. Calculate the percentage of the total land area in your community that is open space.

2. Draw a map that shows your idea of the best way 1 square kilometer of open space can be distributed in a town that has a total area of 10 square kilometers.

3. Conduct a survey to learn which of the possible uses for open space the residents of your community prefer.

Think About This Problem

When a railroad company decided to discontinue a certain branch line, it offered to sell the right of way (about 30 meters wide) to the communities through which the line ran. Some communities refused to buy, claiming that they had no use for the land (Fig. 1-28).

How might this problem affect you?
What caused the problem?
How could you help solve it?

We face problems because of harmful conditions which we ourselves have caused in our environments. To realize how our technology has changed the environment and what we may be able to do to live in greater harmony with nature, we must understand what the environment is like under natural conditions, without the interference of humans. Even though no place on earth has completely escaped the all-pervasive effects of human activities, it is possible to gain some insights into what a world without people might be like. The remainder of this book is concerned with natural conditions and processes in the environment. Chapter 2 is devoted to environmental factors and how they influence living organisms.

Fig. 1-27. How far do you have to walk to get to a recreation area?

(Photo by Alan J. Federow)

Fig. 1-28. This unused railroad right-of-way can make a good bicycle path or hiking trail.

(Photo by Alan J. Federow)

CHAPTER 1. HUMAN ECOLOGY

1. Describe a method of determining whether the slope of land is greater than 10%.

2. Where are floodplains usually located?

3. What harm may be done when swamps or marshes are drained?

4. What is usually done in an attempt to prevent rivers from flooding?

5. How do building construction and other land development increase the danger of flooding?

6. Describe a method of calculating the volume of runoff from impervious surfaces during a 1-cm rainfall.

7. How could you determine the rate at which water flows in a stream?

8. What are some ways of reducing the amount of runoff water that gets into streams?

9. Why can automobile exhaust encourage algae to grow rapidly in a pond?

10. What are some human diseases that may be caused by automobile exhaust?

11. What are some methods of reducing the rate at which automobile exhaust enters the air?

12. An automobile travels ¼ kilometer in 20 seconds. What is its velocity in kilometers per hour?

13. In a certain community, 9200 people live in 3.25 square kilometers. What is the population density of the community?

14. A television set rated at 300 watts is used an average of 6 hours per day. How many KWH of energy does the television set use in a 30-day month?

15. An electric clock is rated at 2 watts. How long could you use a 100-watt lamp bulb on the energy that the clock uses in 24 hours?

16. List some ways in which parks and other kinds of open spaces benefit people.

Chapter 2
ENVIRONMENTAL FACTORS

Man can live in almost any kind of environment even though he is not naturally adapted to some of the habitats in which he lives. Man builds houses, irrigates fields, and does many other things to change environmental conditions and make them suitable for his existence. He can also change his own characteristics by wearing clothing or using weapons, tools, machinery, and instruments so that he can cope with a hostile environment. In this respect, other species of organisms are not so fortunate. Although each species is adapted to the present conditions in its environment, it cannot deliberately do anything to adapt to rapid changes in the environmental conditions. Only through the long, slow process of evolution can it adapt to environmental changes. If environmental conditions change rapidly in a particular habitat, some species will no longer be able to exist there. For example, after a pond has been filled in with soil, mallard ducks cannot live there because they need open water.

An environment consists of biotic factors (living organisms) and abiotic factors (the nonliving materials and conditions that affect the living organisms). The abiotic factors help determine the kinds and numbers of plants and animals that can live in an area and how well they can live and grow. The way plants live and grow has a great deal to do with our own lives. Along with almost all other organisms, we depend on them for food and oxygen. Plants also are the sources of numerous materials we use in construction and manufacturing. By investigating the effects of various environmental factors, you will better understand the environmental conditions that will probably be most beneficial to you and others in your community.

The effects of abiotic factors on green plants are extremely important. Sunlight, for example, supplies the energy for photosynthesis and is therefore essential for the production of the food and oxygen which practically all organisms require. The intensity and duration of sunlight help determine which species of plants can grow in a particular area. If you wish to establish a garden in a place that does not receive full sunlight all day, it is important for you to know which kinds of plants will thrive there. In Project 2-1 you will investigate one way in which light can affect plant growth. In this project you will use artificial light instead of sunlight.

PROJECT 2-1: How light affects plant growth.

You will need: ten soybeans, a beaker, two 3" flowerpots, two labels, potting soil, water, a Grolux bulb in a suitable fixture, a timer for electrical devices, a notebook, graph paper, a pencil, and a pen.

1. Soak the soybeans overnight in a beaker of water.
2. Fill each flowerpot with soil.
3. Plant five soybeans in each flowerpot.
4. Set the timer so that the Grolux lamp is on 14 hours per day. Place one flowerpot, labeled

Environmental Factors

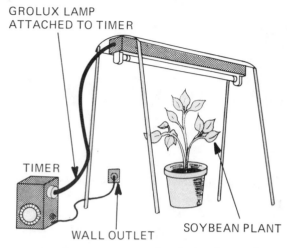

Fig. 2-1. *How to arrange the Grolux lamp, timer, and soybean plant.*

LIGHT, under the Grolux lamp (Fig. 2-1). Place the other flower pot, labeled DARK, in a dark closet.

5. Water the soil in each pot thoroughly immediately after planting; then water thoroughly on every subsequent Monday and Friday.

6. Measure and record the lengths of the plants each Monday. Calculate the average length of the plants in each pot.

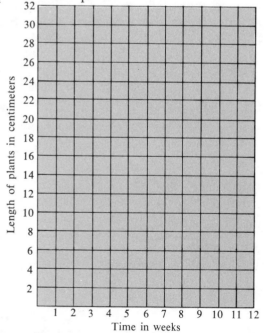

Fig. 2-2. *Graph to show the growth of soybean plants in the light and in the dark.*

7. Make a graph of your results (Fig. 2-2). Draw a separate line on the graph for the plants in each pot.

8. Compare the results for each pot. What is the relationship between light and rate of growth of soybeans?

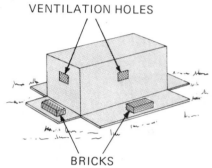

Fig. 2-3. *Cardboard carton inverted on lawn.*

For Further Investigation

1. Cut one ventilation hole (approximately 10 centimeters square) in each of the sides of a large cardboard carton. Set the carton upside down on a lawn with its flaps extending outward. Weight the carton down by putting bricks or stones on the flaps (Fig. 2-3). Do not mow the lawn within 1 meter of the carton. After one week, measure, record, and compare the lengths of the grass both inside and outside the carton.

2. The rapid growth of plants that occurs when they are illuminated by very low intensity light (or not illuminated at all) is known as *etiolation*. Try to find some places where etiolation has occurred naturally or by accident.

3. Of what benefit might etiolation be to a plant?

4. Compare the distance between leaves on etiolated plants and on normal plants.

Think About This Problem

Most plants do not grow well in gardens surrounded by buildings.

How might this problem affect you?
What causes the problem?
How could you help solve the problem?

In Project 2-1 you probably found that those plants that received less light grew faster (became etiolated). If you are wondering why this is so, you may find part of the answer when you perform Project 2-2. Here you will compare the masses of etiolated plants with those of normal plants. You will also determine the ratio of dry mass to gross mass for both etiolated plants and normal plants. The results you obtain will help you decide how the increase in dry mass is related to the intensity of illumination.

PROJECT 2-2: How etiolation affects the ratio of dry mass to gross mass.

You will need: twenty soybeans, a 400-milliliter beaker, two large flowerpots, potting soil, a Grolux bulb in a suitable fixture, a timer for electrical devices, water, two evaporating dishes, a pair of crucible tongs, a chemical balance, a set of masses for the chemical balance, a drying oven, a notebook, and a pen.

1. Soak the soybeans over night in some water in the beaker.
2. Fill each flowerpot with soil. Plant ten soybeans in each pot. Water the soil thoroughly.
3. Put one pot in full sunlight (or under the Grolux bulb, timed to be on 14 hours per day) and place the other pot in a dimly illuminated location. The temperatures that the pots are exposed to should be between about 18° Celsius and 30° Celsius.
4. Water the soil thoroughly each Monday and Friday.
5. At the end of one month, measure and record the lengths of the plants in both flowerpots.
6. Harvest the plants, keeping the two groups separate. Try to remove the entire plant, including the root. To do this, empty the soil, along with the plants, out of the pots; then wash away the soil.
7. Measure and record the mass of each evaporating dish.

Each of the following procedures is to be performed separately for each of the two groups of plants.

8. Crumple the plants and fit them into the evaporating dish. Do not squeeze the plants hard enough to remove any juices.
9. Measure and record the mass of the evaporating dish plus plants.
10. Calculate the mass of the plants by subtracting the mass of the evaporating dish from the mass of the evaporating dish plus plants.
11. Heat the evaporating dish and plants in an oven at 95° Celsius for 1 hour. Remove the evaporating dish from the oven and allow it to cool to approximately room temperature.
12. Measure and record the mass of the evaporating dish and plants.
13. Heat and the evaporating dish in the oven at 95° Celsius for 15 minutes. Remove the evaporating dish and contents from the oven and allow them to cool to approximately room temperature.
14. Measure and record the mass of the evaporating dish plus contents. If there is a difference between the results of Steps 12 and 14, repeat Steps 13 and 14 until there is no further significant change in mass. Then proceed to Step 15. If there is no significant difference between the results of Steps 12 and 14, proceed to Step 15.
15. Calculate and record the dry mass of the plants by subtracting the mass of the evaporating dish from the mass of the evaporating dish plus contents.
16. Calculate the ratio of the dry mass of the plants to their original (gross) mass.
17. Compare the normal and etiolated plants as to their (a) gross masses, (b) dry masses, (c) ratio of dry mass to gross mass.
18. Explain the results.

For Further Investigation

1. Use a microscope with a magnification of about 400× to examine cross sections of the

ENVIRONMENTAL FACTORS

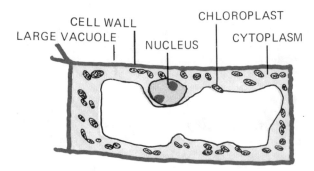

Fig. 2-4. One cell of a soybean leaf as seen under the high power of a microscope.

leaves of normal soybean plants (Fig. 2-4) and leaves of soybean plants that have been grown in the dark. Compare the two with respect to the number of chloroplasts and the size and shape of the cells.

2. "Blanched" asparagus is produced by heaping soil over asparagus shoots as they grow. Why are these plants white and tender?

Think About This Problem

Plants often grow too tall and spindly when they are raised indoors.

How might this problem affect you?
What causes the problem?
How could you help solve it?

Project 2-2 should provide some evidence that greater light intensity causes a greater rate of photosynthesis, even though plants grow more rapidly in length under light of decreased intensity.

It is evident that the degree of intensity of light has great effects on plants. In fact, intensity of sunlight plays a large role in determining which species of plants will grow in a locality. If you know how intense the light must be for various kinds of plants, you may avoid failure when you try to grow them. As you perform Project 2-3, you will find out how light intensity may vary from place to place and how the species of plants also vary.

PROJECT 2-3: How variation in light intensity helps determine the kinds of plants that can grow in an area.

You will need: a metric tape measure, a ruler, 21 wooden stakes, a hammer, a quadrat frame (see Project A-4 in the Appendix: Making a quadrat frame), cord, a light meter, white paper, a notebook, and a pen.

1. Select an area where a field borders on a forest (or a grove of trees). Lay out a transect 20 meters long, going from the field into the forest (Fig. 2-5). (See Project A-3 in the Appendix: How to lay out a transect.)

2. When the sun is shining, measure and record the light intensity at each meter of the transect length. Place the piece of white paper on the ground and measure the illumination with a light meter held about 20 centimeters from the paper (Figs. 2-6 and 2-7).

NOTE: You may use any one of a number of kinds of light meters. A standard scientific light meter calibrated in lumens/meter^2 or in foot candles is preferable. However, a photographer's light meter may be used. You might, for example, set the ASA number at 25, set the time at $1/60$ of a second, and read the "f" stop number on the light meter as a measure of the light intensity.

3. Draw a graph of the light intensity against the distance along the transect (Fig. 2-8).

4. Differentiate and record the kinds of plants in each square along the transect (Fig. 2-9), using the scheme given in Project A-6 in the Appendix: Differentiating plants. In order to carry out this step, place a quadrat frame (see Project A-4 in the Appendix: Making a quadrat frame) on the ground at each meter of the transect and differentiate the plants within the frame.

5. Write out a table that tells which kinds of plants are present at each range of light intensity. You might arrange your table as shown in Fig. 2-10.

6. Explain your results.

Fig. 2-5. Laying out a transect. *(Photo by Alan J. Federow)*

Fig. 2-6. A light meter.
(Courtesy of Berkey Marketing Corp.)

Fig. 2-7. Measuring the illumination of a sheet of paper on the grass with a light meter.
(Photo by Alan J. Federow)

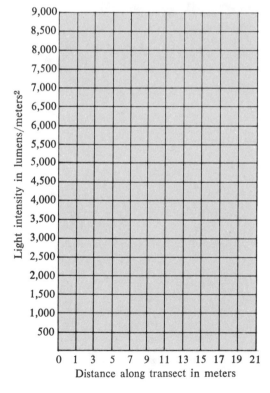

Fig. 2-8. Graph to measure the variation of light intensity along a transect.

Fig. 2-9. Differentiating plants in a quadrat. (Photo by Alan J. Federow)

PLANTS

INTENSITY RANGE IN LUMENS/METERS^2	CODE NUMBER	NAME
0- 250		
250- 500		
500- 1,000		
1,000- 2,000		
2,000- 3,000		
3,000- 4,000		
4,000- 5,000		
5,000- 7,500		

Fig. 2-10. A table listing the plants that are present intensity along a transect.

For Further Investigation

1. Determine the number of individual plants in each quadrat of the type of transect described in Project 2-3. Consider each stem growing from the soil to represent an individual plant. Draw a graph illustrating your findings.

2. Measure and record the variation of light intensity at different seasons of the year along the type of transect described in Project 2-3. Draw appropriate graphs.

3. Measure and record light intensities at several locations in which there are definite differences in the kinds of trees or other tall plants.

Think About This Problem

Park planners sometimes try unsuccessfully to grow a lawn near a group of trees.

How might this problem affect you?
What causes the problem?
How could you help solve it?

Radiant energy from the sun is important to us in many ways. Aside from illuminating the

Fig. 2-11. Catfish (below) can live in thermally polluted streams. Trout (above), a more desirable fish, cannot survive there.
(Courtesy American Museum of Natural History)

earth and providing energy for photosynthesis, the sun warms the earth and its atmosphere. Variation in the intensity of solar radiation that reaches a given area helps determine whether it will be warm or cold and thus helps cause the seasons of the year. In turn, this affects your comfort and the activities you can engage in. Whether you will be ice skating or swimming in a lake depends largely on the outdoor temperature, which depends, to a large extent, on the intensity of solar radiation.

Humans can live almost anywhere on earth, regardless of the outdoor temperature, mainly because they have developed warm clothing and methods of heating their homes. Most other organisms can live only where the temperature remains within certain limits. Fish are especially sensitive to temperature changes. The fresh-water fish (such as trout) that are most valuable as food and for sport fishing generally can live only where water temperature is relatively low. If the water temperature rises significantly (because, for example, industries thermally pollute the water), the trout and other desirable fish are either killed or can no longer reproduce. These fish gradually disappear and are replaced by carp, garfish, catfish, and other fish that are either inferior or useless as food or for sport fishing (Fig. 2-11).

In Project 2-4 you will investigate one way that temperature changes affect fish.

ENVIRONMENTAL FACTORS

PROJECT 2-4: Effect of temperature changes on the breathing rate of goldfish.

You will need: two small live goldfish (about 4 centimeters long), a small fishnet, two 1000-milliliter pyrex beakers, spring water, a 500-milliliter graduated cylinder, an electric hotplate, two thermometers (range from $-10°$ Celsius to $110°$ Celsius), a watch with a second hand, a notebook, and a pen.

1. Using the graduated cylinder, pour about 500 milliliters of cool, well-aerated spring water into each of two 1000-milliliter pyrex beakers. Put a thermometer into each beaker, making certain that the thermometer cannot fall out of the beaker. Put a goldfish into each beaker (Fig. 2-12).
2. Record the water temperature for each beaker.

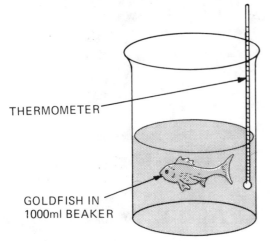

Fig. 2-12. Setup to determine how temperature affects the breathing of goldfish.

Count and record the number of gill movements each goldfish makes in 1 minute.

3. Slowly heat one of the beakers (with the goldfish inside) on an electric hotplate.
4. When the water temperature reaches $25°$ Celsius, record the temperature of the heated water and count and record the number of gill movements in 1 minute. Immediately after doing so, record the temperature of the cool water and count and record the number of gill movements the fish in the cool water makes in 1 minute.
5. Repeat Step 4 when the heated water reaches $30°$ Celsius.
6. Repeat Step 4 when the heated water reaches $35°$ Celsius.

7. Remove the goldfish from the heated water and put it into the beaker of cool water with the other goldfish.

NOTE: Ordinarily, goldfish can live in water at $35°$ Celsius without being harmed. If, however, the goldfish in the heated water makes frantic efforts to breathe air at the surface, it is not receiving enough oxygen. Remove it from the heated water and put it into the cool water.

8. Draw a graph of the results obtained for the fish in the heated water and another graph of the results obtained for the fish in the cool water. Your graphs might be arranged as shown in Fig. 2-13.
9. Explain your results.

For Further Investigation

1. Measure the water temperatures in a stream, lake, or pond. Correlate the water temperature

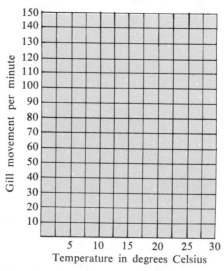

Fig. 2-13. Graph to show how the rate of gill movements varies as the water temperature increases.

with the species of fish present in the body of water.

2. Using a fisherman's temperature measuring device, measure the temperatures in various sections of a lake and at various depths. Correlate the temperatures in each location with the species of fish. Local fishermen may help provide the required information.

Think About This Problem

Only carp, catfish, and garfish can now live in many rivers where there were trout and bass before the rivers were thermally polluted.

How might this problem affect you?
What causes the problem?
How could you help solve it?

The concentration of a dissolved gas that water can hold decreases as the water temperature increases. Warm water holds less of any dissolved gas than an equal quantity of cool water. As water is warmed, some dissolved oxygen (a gas) is driven off, making it more difficult for a fish to obtain the oxygen it needs. To compensate for the decreased concentration of oxygen, the fish breathes faster.

Various species of fish differ in their oxygen requirements. This is one reason that different fish species vary in their ability to live in warm water. Goldfish, a species of carp, have unusual tolerance for high water temperatures and low dissolved oxygen levels. Brook trout, on the other hand, cannot thrive where the water temperature exceeds 20° Celsius.

Thermal pollution (heating) of streams and lakes is fairly common. It is generally caused by hot water flowing from factories and electric power plants. Many of these establishments draw cool water from streams and use it for cooling hot metals, condensing steam, and similar purposes. After the water has become hot, it is allowed to pour back into the stream, raising its temperature. This rise in temperature kills some fish and makes it impossible for others to grow well in the stream.

Most water animals, such as fish, are much more sensitive to temperature changes than most land animals. One reason may be that increased temperature decreases the percentage of oxygen in water but not in air. Still, temperature differences also have fundamental effects on land organisms. We will observe how temperature changes affect a particular land organism in Project 2-5.

PROJECT 2-5: The effect of temperature on the metamorphosis of mealworm pupae.

You will need: ten mealworm larvae of approximately equal size, three 400-milliliter beakers, oatmeal, a potato, a knife, a metric ruler, a refrigerator, a refrigerator thermometer, a notebook, and a pen.

1. Cut a cube of potato that is about 1 centimeter in each dimension. Put the cube of potato into a beaker. Pour oatmeal into the beaker until it is half full (Fig. 2-14).

2. Put the ten mealworm larvae into the beaker prepared in Step 1. Allow them to remain at room temperature until they become pupae (Fig. 2-15). This is called *pupation*. Add new cubes of potato to the beaker if the original cube becomes dry.

3. Set the refrigerator at 5° Celsius. Check the temperature with the thermometer to see if the setting is correct.

4. Prepare two more beakers as you did in Step 1. Put one beaker into the refrigerator. Leave the other beaker on a table or shelf in the room.

5. As the larvae become pupae, put five pupae into each of the two beakers. If fewer than ten pupae develop, put equal numbers of larvae into each beaker. Record the date and time that each

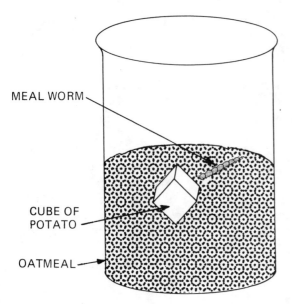

Fig. 2-14. Setup to determine how temperature affects the metamorphosis of a mealworm.

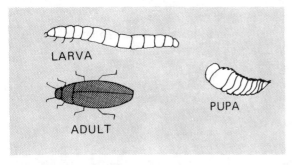

Fig. 2-15. Stages in the development of a mealworm.

pupa is added to a beaker. It is not necessary to mark the pupae or identify the individuals in any way.

6. Observe and record the date (and, if possible, the time) that each pupa becomes an adult beetle: Remove the beetles and put them into the original beaker which was prepared in Step 1.
7. Calculate the average length of time from pupation to the emergence of an adult. Do this separately for the pupae in the room and the pupae in the refrigerator.
8. Explain the results.

For Further Investigation

1. Perform projects similar to Project 2-5, keeping sets of pupae at 10° Celsius and 15° Celsius in a refrigerator and other sets of pupae in an incubator at temperatures of 25°, 30°, and 35° Celsius. Determine the temperature at which the mealworms remain as pupae for the least time before becoming adults.

2. Perform projects similar to Project 2-5 with fruit flies.

Think About This Problem

Noisy airplanes near airports frequently frighten birds from nests where they are incubating eggs. As a result, some of the eggs never hatch and defective birds hatch from some of the other eggs.

How might this problem affect you?
What causes the problem?
How could you help solve it?

In Project 2-5 we investigated the influence of temperature on one species of animal. Does temperature also have important effects on plants? Let's find out in Project 2-6.

PROJECT 2-6: How temperature affects sprouting of seeds.

You will need: 20 bush bean seeds, three 400-milliliter beakers, blotting paper, water, a refrigerator, a refrigerator thermometer, a notebook, and a pen.

1. Soak the bean seeds overnight in water in one of the beakers.
2. Put a piece of blotting paper into each of the two remaining beakers so that it fits against the side. Pour water into each beaker until the blotting paper is completely soaked and about 1 centimeter of water remains in the bottom of the beaker.
3. Place ten bean seeds between the blotter and the side of each beaker (Fig. 2-16).

ENVIRONMENTAL FACTORS

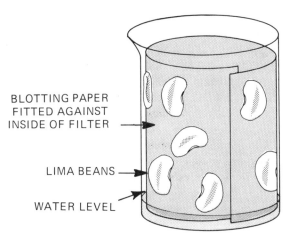

Fig. 2-16. Setup to determine how temperature affects the sprouting of seeds.

4. Set the refrigerator temperature at 10° Celsius. Check the temperature with the thermometer.
5. Put one beaker with bean seeds in the refrigerator. Leave the other beaker on a table or shelf in the room. Record the date and time that this is done. Measure and record the temperature in the room.
6. Observe and record in your notebook the date that each of the seeds sprout.
7. Calculate the average time required for the sprouting of each of the two sets of seeds, one in the refrigerator and one in the room. If seeds have not sprouted within 3 weeks, assume that they will never sprout.

For Further Investigation

1. Perform projects similar to Project 2-6 with sets of bean seeds in a refrigerator at 15° Celsius and in an incubator at 25°, 30°, 35°, and 40° Celsius. Determine the temperature at which sprouting takes the least time.

2. Once a week beginning in early spring and continuing for at least 2 months, measure the temperature of the soil at a depth of about 3 centimeters below the surface. If possible, leave the thermometer in place for the entire period during which you are measuring the temperature. Graph your results. Correlate your findings with the time seeds are generally planted in the locality.

3. *Herbaceous* plants are those that do not have woody stems or trunks. Beginning in early spring, observe a lawn, field, or empty lot, noting when the first new, green shoots of the different species of herbaceous plants appear above ground. When you find such a shoot, dig it out of the ground along with a small quantity of soil. Wash this soil away. Note the nature of the underground part of the plant. It may consist of a bulb, rhizome, thin roots, thick roots, or taproot (Fig. 2-17).

Which kind of underground part indicates that the plant sprouted from a seed just a few days previously? Record the dates when you discover new shoots and the type of underground part each has. Determine whether sprouts from seeds appear before or after shoots of other plants.

Measure the soil temperature in the area to determine the temperatures at which seeds of wild plants first sprout.

Think About This Problem

In some parts of the country, the growing season is too short for growing certain plants such as melons. When gardeners attempt to lengthen the season by planting earlier, the seeds do not sprout.

How might this problem affect you?
What causes the problem?
How could you help solve it?

You have found that temperature helps determine whether or not seeds will sprout. Temperature also affects the opening of flower and leaf buds on trees and shrubs. In warmer climates

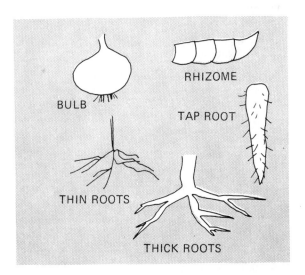

Fig. 2-17. Several kinds of underground plant parts.

ENVIRONMENTAL FACTORS

the buds open earlier, the trees and shrubs grow more during the growing season, and fruits ripen earlier than in cooler climates. As a result, farmers in warmer climates have some advantages over those in cooler climates. Farmers fully realize the importance of temperature for rapid, sturdy plant growth. To provide the proper temperatures for certain plants, some farmers use hothouses, greenhouses, or coldframes (which are like small greenhouses). Without resorting to such artificial devices, farmers (and you yourself) can locate places within your general area where certain plants will thrive because the temperature and other weather factors are more suitable than in other places.

Microclimates are the differences in the weather factors at various locations within a relatively small area (anything up to a few hectares). Microclimates are important because of the way the different weather factors affect both plants and animals. In Project 2-7, you will investigate one phase of microclimates that is extremely important to anyone who grows plants or is interested in knowing why certain wild plants grow in one particular place and not another in the same general locality.

PROJECT 2-7: Temperature difference between a north-facing slope and a south-facing slope.

You will need: two thermometers, each with a range from about −10° Celsius to 110° Celsius; two cardboard tubes, each about 2 centimeters in diameter and 10 centimeters long; aluminum foil; two rubber bands; a watch; a notebook; and a pen.

1. Wrap each cardboard tube in aluminum foil and fasten a rubber band around it as in Fig. 2-18.
2. Put a thermometer in each tube so that the bulb of the thermometer is inside the tube.
3. Find a place where a north-facing slope and a south-facing slope are fairly close to each other. The most definite results will be obtained with steep slopes (at least 30° from the horizontal). Suitable types of locations include a ridge, a ditch, and the banks of a road-cut. Be sure that the places chosen will not be shaded by trees, shrubs, or other tall objects. If necessary, you may dig a V-shaped ditch about 1 meter wide, 1 meter long, and 50 centimeters deep.
4. Place one tube and thermometer on the north-facing slope and the other tube and thermometer on the south-facing slope (Fig. 2-19).
5. After 30 minutes, read and record the temperature on each thermometer. Leave the tubes and thermometers on the slopes until you have completed the project.
6. Read and record the temperatures several times during the day, at intervals of about 1 hour.
7. Graph the results (Fig. 2-20).

Fig. 2-18. Thermometer in tube, wrapped in aluminum foil.

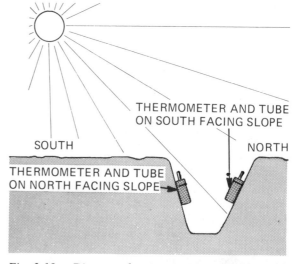

Fig. 2-19. Diagram showing positions of thermometers in ditch.

ENVIRONMENTAL FACTORS

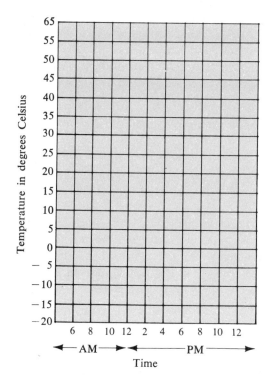

Fig. 2-20. Graph showing temperature variation during the day on the north-facing and south-facing slopes of a ditch.

8. Explain the results.
9. What practical applications are there to our daily lives?

For Further Investigation

1. Perform a project similar to Project 2-7, to compare the temperature under a tree with the temperature on a lawn or field.

2. Design and perform a project to compare the temperatures at various heights (say, 50, 100, 150, and 200 centimeters) above the ground with the temperature at ground level.

3. Design and perform a project to compare the temperature changes during the day on a valley floor with those that occur on the slopes of the adjacent hillsides. If at all possible, measure the changes that take place in the hour or two after sunrise and in the hour or two after sunset.

4. Measure, record, and compare wind velocity in various locations. You will need anemometers of some sort to perform this project.

Think About This Problem

A ski slope on one side of a mountain has a long season and is highly profitable. On the other side of the mountain, there is a ski slope that is unprofitable because its season is several weeks shorter than that of the other ski slope.

How might this problem affect you?
What causes the problem?
How could you help solve it?

In the most familiar kind of desert, which is known as the *hot desert*, there is plenty of strong sunlight and a temperature range suitable for many species of plants, yet few plants may be noticeable during most of the year. After a sudden, drenching rainfall, however, a variety of plants in the same desert may grow rapidly and burst into bloom. The water made the difference. Water in the soil is vital for the growth of plants. This is why there are sprinklers on most lawns and why most farmers are concerned about the amount of rainfall their crops receive. Rain is important to everyone, because it is important to plants. For example, the amount of rain trees receive each year helps determine their rate of growth. This in turn, affects the amount of lumber available for building homes and for other purposes (Fig. 2-21). In Project 2-8, you will investigate the effect of rainfall on the growth of trees.

Fig. 2-21. We need a plentiful supply of lumber for the construction of homes. (Courtesy of N.F.P.A.)

ENVIRONMENTAL FACTORS

> **PROJECT 2-8:** How do variations in annual rainfall affect the rate of growth of trees?

You will need: a cross section of a tree trunk or branch (with the date when the most recent annual ring grew), a metric ruler, a record of annual local rainfall (for the past 50 years) for the region where the tree grew, a notebook, and a pen.

1. Measure and record the width of each annual ring on the cross section of a tree trunk or branch (Fig. 2-22).

Fig. 2-22. Measuring the annual rings of the cross section of a tree trunk.

2. Choose a year during which one of the annual rings grew on your cross section. Divide the widths of each annual ring by the width of the annual ring that grew during the chosen year. We may call the results the ANNUAL RING QUOTIENTS.
3. From the record of annual rainfall, copy the number of inches of rain that fell during each year in your locality when the annual rings grew on your cross section.
4. Divide the number of inches of rain that fell each year by the number of inches of rain that fell during the year chosen in Step 2. We may call the results the RAINFALL QUOTIENTS (Fig. 2-23).
5. Compare the annual ring quotients to the rainfall quotients. Find out whether there is a direct relationship between the widths of the annual rings and the annual rainfall during the years that the annual rings grew.
6. Explain your results.

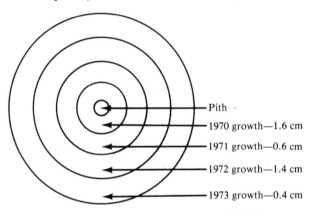

Sizes of annual rings—the chosen year is 1971

YEAR	ANNUAL RING QUOTIENT	ANNUAL RAINFALL	RAINFALL QUOTIENT
1970	$\frac{1.6 \text{ cm}}{0.6 \text{ cm}} = 2.67$	122 cm	$\frac{122 \text{ cm}}{89 \text{ cm}} = 1.37$
1971	$\frac{0.6 \text{ cm}}{0.6 \text{ cm}} = 1.00$	89 cm	$\frac{89 \text{ cm}}{89 \text{ cm}} = 1.00$
1972	$\frac{1.4 \text{ cm}}{0.6 \text{ cm}} = 2.33$	104 cm	$\frac{104 \text{ cm}}{89 \text{ cm}} = 1.17$
1973	$\frac{0.4 \text{ cm}}{0.6 \text{ cm}} = 0.67$	83 cm	$\frac{83 \text{ cm}}{89 \text{ cm}} = 0.93$

Fig. 2-23. Rainfall quotients.

For Further Investigation

1. Compare the annual ring quotients calculated for one cross section with those calculated for an-

other cross section. Determine whether the annual rings on the two cross sections show the same growth pattern.

2. On pine trees a new whorl of branches begins growing each year. The distance between a whorl of branches and the whorl below it represents 1 year's growth. The most recent year's growth is at the top of the tree; the previous year's growth is the one below the top; etc. (Fig. 2-24).

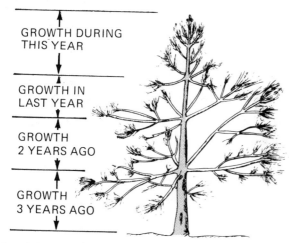

Fig. 2-24. Pine tree showing whorls of new branches that grow every year.

Measure and record the length of each year's growth on a pine tree. Using a procedure similar to that in Project 2-8, compare the variation in growth to the variation in annual rainfall in the years during which the tree grew.

3. Measure the mass of a sample of soil, heat the soil to dryness at 95° Celsius, then measure the mass of the soil again. The difference in mass is the mass of water that was in the soil. Use the data you obtain to determine the percentage of water in the soil.

Think About This Problem

In the Sahel, a region in Africa, the inhabitants used to be able to raise all the food they needed. For the past few years, the Sahel has been becoming more like a desert; there have been repeated crop failures and people are dying of starvation.

How might this problem affect you?
What causes the problem?
How could you help solve it?

Although there is a large amount of water on the earth, the supply is still limited. Yet there seems to be no limit to the amount of rain that can fall. Each year an average of about 1 meter of rain falls on all parts of the earth. This has been going on for most of the 4.5 billion years of the earth's existence. The rate of rainfall varies but the average rainfall over all the years has probably not been greatly different from what it now is. It has been possible for rain to keep falling only because it is part of the water cycle. This is a process by which the same water repeatedly passes through different phases and through different parts of the environment to complete a natural cycle again and again.

Natural cycles make essential substances available to living plant and animal organisms over and over again. The same molecule of water may have become part of a plant or animal many thousands of times. There are natural cycles for a number of substances in addition to water. You will investigate some aspects of these cycles in Chapter 3. You will also learn how organisms obtain their energy and how energy is transferred from organism to organism.

CHAPTER 2. ENVIRONMENTAL FACTORS

1. How may rapid environmental changes affect various species of organisms?

2. What are some kinds of abiotic environmental factors?

3. What causes etiolation?

4. How can you determine the dry mass of a plant?

5. Why may etiolated plants have a relatively low ratio of dry mass to gross mass?

6. What are some reasons why the plants under the trees in a forest are different from the plants in a field?

7. How can you measure the intensity of illumination?

8. What are some kinds of fish found in the temperate zone that can withstand relatively high water temperatures?

9. How is the breathing rate of a fish affected by variation in water temperature?

10. How does variation in water temperature affect the amount of dissolved oxygen the water can hold?

11. What are the main causes of thermal pollution of water?

12. What are the stages in the metamorphosis of the mealworm?

13. How does soil temperature affect the sprouting of seeds?

14. What are microclimates?

15. What is the water cycle?

Chapter 3

NATURAL CYCLES AND TRANSFER OF ENERGY

Natural cycles can be a matter of life and death. In parts of Africa people are starving because their land is gradually becoming desert. The water cycle there no longer acts rapidly enough to bring the amounts of rain the farmers' crops require.

After precipitation of rain or snow from clouds, the water percolates through the soil, moving slowly downward under the influence of gravity. Water can also move through the soil by capillarity, a process in which water is pulled upward (or in any other direction) by the force of surface tension through the small spaces between soil particles. Root hairs, extremely thin projections from cells near the tips of roots, push their way into spaces and absorb the water. Only a small fraction of the absorbed water is used in the plant for photosynthesis. Plants release most of the water into the air in almost pure form by the process known as transpiration, after extracting the minerals the water contained. Huge quantities of water enter the atmosphere by transpiration.

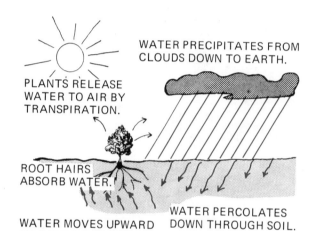

Fig. 3-1. Transpiration.

Water also enters the atmosphere by evaporation of water directly from the soil or from bodies of water (Fig. 3-1).

You will investigate the way the leaves of a plant affect the rate of transpiration in Project 3-1.

PROJECT 3-1: How the leaves of a plant affect the rate of transpiration.

You will need: two twigs (from the same tree or shrub) each about 30 centimeters long and with a diameter of about 7 millimeters at the base and bearing leaves; two gas-collecting bottles; two two-holed rubber stoppers to fit the gas-collecting bottles, each stopper fitted with an an L-shaped piece of glass tubing (7 millimeters in diameter, each arm about 10 centimeters long); two pieces of rubber tubing (each about 10 centimeters long) to fit the glass tubing; two pieces of capillary tubing (each about 20 centimeters long) with a 7-millimeter outside diameter; a medicine dropper; water paraffin wax; a pan; an electric hotplate; a small watercolor brush; a watch; a ruler; a notebook; and a pen.

1. Remove all the leaves from one twig.
2. Pour water into each collecting bottle until it is about three-fourths full.
3. Fit a rubber stopper, with L-shaped glass tube, into each bottle. Attach the rubber tubing and glass capillary tubing as shown in Fig. 3-2.

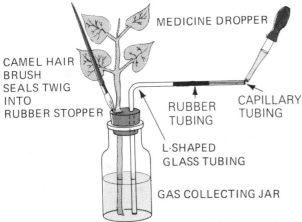

Fig. 3-2. Setup to study how a plant's leaves affect its rate of transpiration.

4. Insert each twig into the empty hole in one of the rubber stoppers. Push the twig into the hole until the end of the twig is just above the bottom of the bottle.
5. Heat a small quantity of wax in a pan on an electric hotplate until the wax melts. Stop heating the wax as soon as possible after it has melted.
6. Using the melted wax, seal each twig into the hole in the rubber stopper (Fig. 3-2).
7. Pick up some water with the medicine dropper. Introduce a drop of water into the open end of each glass capillary tube (Fig. 3-2).
8. Measure and record the distance that each water drop moves in 5 minutes. This distance is a measure of the amount of water that has been removed from the bottle and passed through the plant by transpiration.
9. Remove half of the leaves from the twig that still bears leaves. Repeat Steps 7 and 8.
10. Remove half of the remaining leaves. Repeat Steps 7 and 8.
11. Explain your results.

For Further Investigation

1. Following a procedure similar to that in Project 3-1, compare the rates of transpiration of equally long twigs from a maple tree and from a pine tree.
2. Design and perform an investigation to measure the number of milliliters of water that a twig transpires in 15 minutes.
3. Coat the leaves of a twig with petrolatum (Vaseline); then measure its rate of transpiration.

Think About This Problem

After a tree was transplanted, its leaves shriveled and the tree died.

How might this problem affect you?
What caused the problem?
How could you help solve it?

Plants wilt or even die if their roots cannot absorb water rapidly enough from the soil to make up for transpiration losses. In the case of typical land plants, transpiration takes place almost entirely through the leaves. Pine trees and other conifers have relatively low rates of transpiration because their needlelike leaves have waxy coatings and also expose little surface area to the air. These adaptations help the conifers survive the winter when most of the water in the soil is not available because it is frozen. The deciduous trees reduce transpiration to very low levels in the winter by simply dropping their leaves. Some desert plants, such as cacti, have no leaves at any time. Other desert plants may lose their leaves during droughts and sprout new leaves after a rainfall.

Transpiration presents great problems to desert plants because the soil contains little water and the air is so dry. When their soils are supplied with enough water, deserts often prove to be highly fertile. Irrigation has made it possible for plant life to flourish luxuriantly in parts of the southwestern United States, in Israel, and other areas which previously were forbidding deserts. As a result, such regions have become pleasant places for those people who live there. They also provide additional food for people in many other parts of the world. Irrigation has conquered the problems resulting from low relative humidity and insufficient water in the soil. Project 3-2 examines the effect of relative humidity of the air on the rate of transpiration.

PROJECT 3-2: How relative humidity affects the rate of transpiration.

You will need: the same equipment as in Project 3-1, a sheet of clear plastic (about 1 meter square), a corrugated cardboard carton (about 50 centimeters in each dimension), a metric ruler, a pencil, a knife, a stapler, a cellulose sponge, a plate to hold the sponge, a notebook, a pen, and equipment for measuring relative humidity (see Project A-15 in the Appendix: How to measure relative humidity).

1. Cut the flaps off the cardboard carton. Cut large rectangular openings in the top and sides of the carton. Leave enough of the cardboard to serve as a support for the plastic.
2. Cover the top and sides of the carton with plastic and staple the plastic in place (Fig. 3-3).
3. Soak the sponge in water and place it in the plate. Put the plate and sponge on the table or other surface where the project will be conducted and invert the plastic-covered carton over them (Fig. 3-4).
4. Pour water into each collecting bottle until it is about three-fourths full.
5. Fit a rubber stopper (with L-shaped glass tube) into each bottle. Attach the rubber tubing and glass capillary tubing as shown in Fig. 3-2.
6. Insert each twig into the empty hole in one of the rubber stoppers. Push the twig into the hole until the end of the twig is just above the bottom of the bottle.
7. Heat a small quantity of wax in a pan on an electric hotplate until the wax melts. Stop heating the wax as soon as possible after it has melted.
8. Using the melted wax, seal each twig into the hole in the rubber stopper.
9. Pick up some water with the medicine dropper. Introduce a drop of water into the open end of each glass tube (Fig. 3-2).

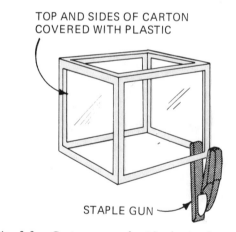

Fig. 3-3. Carton covered with plastic sheet.

Fig. 3-4. Carton inverted over water-soaked sponge.

10. Put one setup under the plastic-covered carton. Put the other setup on the work surface next to the cardboard carton.

11. Measure and record the relative humidity under the plastic-covered cardboard carton and also next to the cardboard carton.

12. Measure and record the distance the water drop moves in 5 minutes in the horizontal capillary glass tube of each transpiration setup. This distance is a measure of the amount of water that has been removed from the bottle and passed through the plant by transpiration.

13. Describe and explain the effect of relative humidity on the rate of transpiration.

For Further Investigation

1. Repeat Project 3-2 in places where there are natural differences in relative humidity, for example, in a forest and in an open field. Shade the setup and protect it from wind to ensure that relative humidity is the only variable factor.

2. Design and perform a project to determine the effect of intensity of illumination on rate of transpiration.

3. Design and perform a project to determine the effect of temperature on rate of transpiration.

Think About This Problem

Some people would not consider living in Arizona because few kinds of trees can grow there unless the soil is watered frequently and thoroughly.

How might this problem affect you?
What causes the problem?
How could you help solve it?

Although both an awning and a tree can shade you from the sun, it may be swelteringly hot under an awning but delightfully cool under a nearby tree. The difference depends on transpiration, the processes by which water vapor escapes into the air through the stomates in a leaf. When water is transpired, it changes from the liquid phase to the gaseous phase (water vapor). This change requires energy, which the plant absorbs from sunlight. Most of the solar energy the plant receives is used for vaporizing water. As the water evaporates, it carries off the solar energy, thus preventing the plant from becoming as hot as an object that does not transpire. You can try to confirm this conclusion by performing Project 3-3.

PROJECT 3-3: Why it is cool under a tree.

You will need: a plastic tree, a thermometer, a notebook, a watch, and a pen.

1. Find a broad-leaved tree with its lower leaves within reach. Place the plastic tree in full sunlight near the real tree.

2. Feel the lower surfaces of the leaves of the plastic tree and the real tree to judge their relative temperatures by using your own temperature sense.

3. Explain the results.

For Further Investigation

1. Put a green Astroturf mat on a lawn in full sunlight. After about 15 minutes, feel both the mat and the lawn. Which is cooler?

2. Compare the temperatures under a tent and under a tree when both are in full sunlight.

Think About This Problem

Even in the shade of buildings, a city street is usually much hotter in the summer than a field or forest in the same general area.

How might this problem affect you?
What causes the problem?
How could you help solve it?

Almost all the water that a plant absorbs from the soil leaves the plant during transpiration, unchanged in composition. Only a small percentage of the water is used by the plant in the process of photosynthesis. Hydrogen from this water becomes part of glucose (a sugar produced in the photosynthetic process) while the oxygen from the water becomes gaseous oxygen, some of which is released into the atmosphere.

The production of glucose requires carbon dioxide in addition to the hydrogen that water supplies. Although there are various natural sources of carbon dioxide, this gas is released at the greatest rate by the process of cellular respiration which occurs in both plants and animals. In Project 3-4 you will investigate respiration in a goldfish and its relationship to carbon dioxide and oxygen.

PROJECT 3-4: How an animal may affect the carbon dioxide–oxygen cycle.

You will need: two clear glass jars (each with about 500-milliliter capacity) with covers, distilled water, a 500-milliliter graduated cylinder, bromthymol blue solution, a medicine dropper, a stirring rod, a soda straw, two labels, a small goldfish (about 6 centimeters long), a watch, a notebook, and a pen.

1. Pour 400 milliliters of distilled water into one of the glass jars (Jar #1).
2. Add 5 drops of bromthymol blue solution to the water in the jar; then stir the water. If the water has an easily discernible blue color, proceed to Step 3. If not, add more bromthymol blue solution, 5 drops at a time until a clearly visible blue color results. Now proceed to Step 3.

3. Pour half the water from Jar #1 into the other jar (Jar #2).
4. Put the goldfish into Jar #1. Record the time at which you do so.
5. Cover both jars tightly (Fig. 3-5).
6. Watch for any color change.
   a. In which jar (or jars) does the color change?
   b. Describe the color change.
   c. How long did it take before the color change first became noticeable?
   d. How much time elapsed before the final color change took place?
7. Use the soda straw to exhale through the water in Jar #2. Describe the results.
8. Explain your results.
9. Remove the goldfish from Jar #1 and put it back where it was before the beginning of this project.

For Further Investigation

1. Repeat Project 3-4 but tap on Jar #1 frequently to make the goldfish swim about rapidly.
2. Find out why the remains of plants and animals deep below the surfaces of stagnant bogs do not decay.

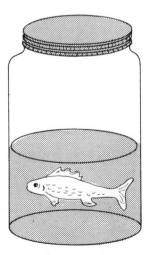

Fig. 3-5. Goldfish in water to which bromthymol blue has been added.

57

NATURAL CYCLES AND TRANSFER OF ENERGY

Fig. 3-6. No fish can live in this stream because the percentage of dissolved oxygen has been reduced too much by the sewage flowing into it.   (Courtesy American Museum of Natural History)

Think About This Problem

The percentage of dissolved oxygen is so low in streams into which untreated sewage is permitted to flow that no fish can live in these streams (Fig. 3-6).

How might this problem affect you?
What causes the problem?
How could you help solve it?

As a goldfish (or any other familiar organism) respires, it uses up oxygen and produces carbon dioxide. This is a major part of the carbon dioxide–oxygen cycle. In another main division of this cycle, carbon dioxide is consumed and oxygen is produced. The carbon dioxide–oxygen cycle is vitally important to most organisms. You could not live for more than a few minutes without oxygen. What supplies the oxygen almost all organisms require? This is the question you will investigate in Project 3-5.

PROJECT 3-5: How oxygen is produced during photosynthesis.

You will need: a 1000-milliliter beaker, distilled water, sodium bicarbonate, a stirring rod, a glass funnel, a 13-millimeter × 100-millimeter test tube, elodea, white thread, a pair of scissors, a wooden splint, matches, a notebook, and a pen.

1. Pour distilled water into the beaker until it is about three-fourths full.
2. Add a large pinch of sodium bicarbonate to the distilled water in the beaker. Stir the water with a stirring rod until the sodium bicarbonate dissolves.
3. Cut five pieces of elodea about the length of the side of the funnel.
4. Tie the pieces of elodea loosely together with the white thread.
5. Holding the elodea under the water with the cut ends upward, place the funnel over the elodea, and allow the funnel to rest on the bottom of the beaker. The tip of the funnel should be below the surface of the water in the beaker.
6. Fill the test tube with distilled water. Cover the mouth of the test tube with your thumb; then invert it (Fig. 3-7). Without removing your thumb from the mouth of the test tube, submerge the mouth of the test tube below the surface of the water in the beaker. Now remove your thumb from the mouth of the test tube. Place the test tube over the stem of the funnel without allowing any water to escape from the test tube.
7. Place the setup in full sunlight. Observe the bubbles rising from the cut ends of the elodea stems (Fig. 3-8). Permit the elodea to remain in full sunlight until all the water has been forced from the test tube by the gas that rises from the elodea.
8. Light a wooden splint. After it has been burning for a few seconds, blow out the flame, so

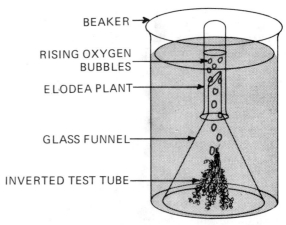

Fig. 3-8. Test tube placed over stem of funnel containing elodea plants.

that a glowing ash remains at the end of the splint.
9. Remove the test tube from the stem of the funnel, covering the mouth of the test tube with your thumb *before* you remove it entirely from the water.
10. When the test tube is out of the water, turn it so that its mouth points slightly upward. Remove your thumb from the test tube. Put the glowing end of the wooden splint into the test tube (Fig. 3-9).
11. Observe and explain the result.

For Further Investigation

1. Find out how intensity of illumination affects the rate of production of oxygen during photosynthesis. Measure the rate of oxygen production by counting the number of bubbles produced per minute. Vary the intensity of illumination by placing the elodea at different distances from a Grolux bulb.

2. Repeat the previous project but do not add sodium bicarbonate to the water. What is the function of the sodium bicarbonate?

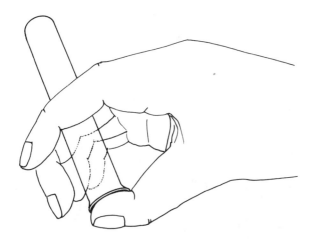

Fig. 3-7. Holding inverted test tube full of distilled water.

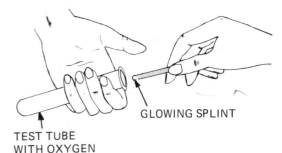

Fig. 3-9. Testing for oxygen with glowing splint.

Think About This Problem

The layer of algae that covers the surface of a eutrophied lake prevents photosynthesis from taking place below the uppermost few centimeters of water.

How might this problem affect you?
What causes the problem?
How could you help solve it?

During photosynthesis green plants (and some other photosynthetic organisms) produce the oxygen that practically all organisms require for respiration, the process by which they obtain energy from food. By the process of photosynthesis, green plants also produce the food that all organisms use for both respiration and growth.

Most of the food that an animal eats is used for energy production by means of the process of respiration. Only a fraction is used for growth. This helps explain why a piece of land will provide far less food for humans if it is used for raising cattle (or other food animals) than it will if food plants for humans (such as wheat) are grown there. When a person eats wheat, he obtains all the material and energy the wheat contains. If, instead, the wheat is fed to cattle, they "waste" most of the food by using it for their own energy requirements. Only a small fraction of the food contributes to their growth and produces beef that people can eat. Obviously, we can provide food for more people if we use land for raising food *plants* instead of food *animals*. As a first step in estimating how many more people can be fed in this way, we should know how many kilograms of plants an herbivore must eat to increase its own mass by 1 kilogram. In Project 3-6 you will determine the percentage of the mass of a plant that becomes additional mass of the herbivore that feeds on the plant. In this case, the herbivore is a leaf miner, a larva that makes a tunnel in a leaf as it feeds on the leaf tissue.

PROJECT 3-6: How is an animal's increase in mass related to the mass of the food it eats?

You will need: a photographic slide of a leaf containing a leaf miner that is about ready to become a pupa, a slide projector, graph paper, Scotch tape, a pencil, a ruler, a notebook, and a pen.

1. Tape a sheet of graph paper to a wall or other smooth, solid vertical surface.
2. Connect the projector to an outlet. Put the slide into the projector and project it onto the graph paper. Focus until you have a sharp image on the graph paper.
3. Using a pencil, trace the outlines of the leaf miner larva and the tunnel that it made by eating part of the leaf (Fig. 3-10).
4. Remove the graph paper from the surface to which it was taped. Using the ruler, measure and record the length and width of the larva (Fig. 3-11).
5. Measure and record the length of the tunnel.
6. Measure and record the maximum width of the tunnel (Fig. 3-12).

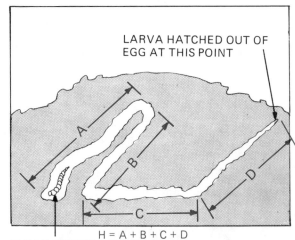

Fig. 3-10. Tracing the outline of leaf miner and tunnel from projected slide.

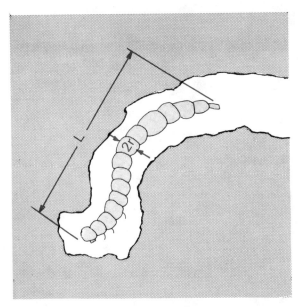

Fig. 3-11. Enlarged drawing showing how to measure length (L) and width (2r) of larva. The radius (r) of the larva is one-half of its width.

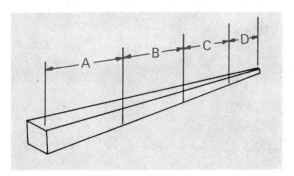

Fig. 3-12. Measuring the dimensions of a leaf miner tunnel.

7. To calculate and record the volume of the leaf miner larva, consider it as a cylinder. Use the formula V = πr²1, where V = volume of the larva, π = 3.14, r = radius (one-half the width of the larva), 1 = the length of the larva.

8. Calculate and record the volume of the tunnel, assuming it to be a triangular prism. Use the formula V = ½ bhw, where V = volume of the tunnel, b = maximum width of the tunnel, h = length of the tunnel, and w = thickness of the tunnel (which we may assume to be equal to the width of the larva).

9. Divide the volume of the tunnel into the volume of the larva and multiply the result by 100. This procedure expresses, as a percentage, the ratio of the volume of the larva to the volume of the tunnel. The ratio of the mass of the larva to the mass of the food it ate may also be represented as the same percentage, provided that we assume that the larva and the leaf have the same density.

For Further Investigation

1. Design and perform a project to determine the percentage of the food eaten by a mealworm larva that becomes additional mass of the mealworm.

2. Design and perform a project to determine the percentage of a kitten's food that becomes additional mass of the kitten. This sort of project can be performed with other kinds of young animals such as white rats or mice.

3. To help provide food for the starving people in parts of our world, it has been suggested that people in the more affluent nations eat a vegetarian diet at least one day per week. Others urge that people in the affluent nations do without at least one meal per week. Which procedure would make more food available to starving people? Explain your answer.

Think About This Problem

About 25% of the crops grown in the United States are eaten by insects (Fig. 3-13).

How might this problem affect you?
What causes the problem?
How could you help solve it?

An average of about 10% of the mass of plants that an herbivore eats is transformed into additional mass of the herbivore. The remainder of the food is either used for respiration or is simply eliminated by the herbivore as waste. By eating the plants, the herbivore obtains the chemical substances that it needs for energy and growth.

Plants contain a variety of nutrients. The sugars, starches, fats, and cellulose are composed of only three elements: carbon, hydrogen, and oxygen. In addition to these three elements, the proteins contain the so-called mineral elements: nitrogen and sulfur. Other mineral elements, including calcium, phosphorus, potassium, sodium, magnesium, and iron, are also found in plant tissues. Every species of organism must take in

Fig. 3-13. Potato bug beetles damage potato crops by eating the leaves. Note the eggs which will soon develop into more potato bugs. (*Courtesy American Museum of Natural History*)

mineral elements to produce proteins, hormones, enzymes, and other substances that comprise its tissues.

It is very difficult to analyze plant tissues to determine the kinds and amounts of elements they contain. However, it is quite simple to find out the total mass of the mineral elements. That is what you will do by performing Project 3-7.

PROJECT 3-7: What is the percentage of minerals in leaves?

You will need: leaves, a pair of scissors, a crucible with a cover, a sensitive balance, masses for the balance, a drying oven, a tripod, a clay triangle, a pair of crucible tongs, a Bunsen burner, a chemical laboratory hood (if available), an asbestos pad, a notebook, and a pen.

1. Using the sensitive balance and masses, measure and record the mass of the crucible.

2. Using the scissors, cut the leaves into pieces small enough to fit into the crucible. Put pieces of leaves into the crucible until it is about three-fourths full. Tap the crucible *gently* against the tabletop to make the leaves settle into place but do not pack them tightly into the crucible.

3. Measure and record the mass of the crucible plus leaves.
4. Subtract the mass of the crucible from the mass of the crucible plus leaves to determine the mass of the leaves. Record the mass of the leaves.
5. Heat the crucible and leaves in the drying oven at about 95° Celsius for 1 hour. With the crucible tongs, remove the crucible from the oven and set it on the asbestos pad to cool.
6. When the crucible and leaves are cool, measure and record their mass.
7. Return the crucible and leaves to the drying oven and heat them for an additional 15 minutes at about 95° Celsius. Remove the crucible from the oven with the crucible tongs and set it on the asbestos pad to cool.
8. When the crucible and leaves are cool, measure and record their mass. If there is a significant difference between this mass and the mass obtained in Step 6, repeat Steps 7 and 8. If there is no significant difference between this mass and the mass obtained in Step 6, proceed to Step 9. Step 8 is to be repeated if any significant difference still remains.
9. Subtract the mass of the crucible from the mass of the crucible plus leaves to obtain the dry mass of the leaves. Record this result.
10. Subtract the dry mass of the leaves from the original mass of the leaves (obtained in Step 4) to determine the mass of water that the leaves originally contained.
11. Measure and record the mass of the crucible cover. Cover the crucible.
12. Place the covered crucible, with leaves, on a clay triangle on a tripod. Heat the crucible with a Bunsen burner for 15 minutes (Fig. 3-14). Perform this step under a laboratory hood or in a well-ventilated room. Turn off the bunsen flame and allow the crucible to cool.
13. When the crucible is cool, measure and record the mass of the covered crucible plus the remains of the leaves.
14. Repeat Steps 12 and 13. If there is a significant difference between the mass obtained in this step and Step 13, repeat Steps 12 and 13 again, until no significant difference remains. If there is no significant difference between the mass obtained in this step and Step 13, proceed to Step 15.
15. Subtract the mass of the crucible and cover from the mass of the crucible, cover, and the remains of the leaves to determine the mass of the remains of the leaves. We may assume that the remains of the leaves consist of mineral matter. This is the only part that does not either evaporate or become gaseous wastes when the leaves are strongly heated.
16. Calculate and record the percentage of mineral matter in dry leaves.
17. Calculate and record the percentage of mineral matter contained in the leaves in their original state.

For Further Investigation

1. Perform Project 3-7 using leaves from different species of plants.
2. Analyze the ash remaining in the crucible qualitatively and quantitatively. Students in advanced chemistry classes in high school or college might assist with this project.
3. Grow a batch of grass in poor soil and another batch in well-fertilized soil. Perform Project 3-7 on each batch of grass and compare the results.
4. Perform Project 3-7 using a piece of meat instead of leaves. It is especially important that this project be performed under a laboratory hood or possibly out of doors since the meat will

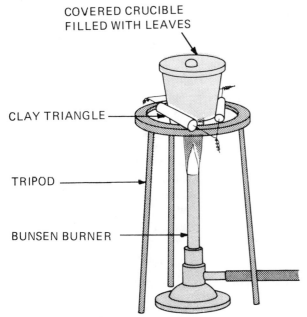

Fig. 3-14. Setup to heat leaves in crucible.

produce very unpleasant odors when it is heated strongly over the Bunsen flame.

Think About This Problem

Enormous amounts of mineral matter are wasted when food scraps are thrown out as garbage.

How might this problem affect you?
What causes the problem?
How could you help solve it?

The fact that plants contain nitrogen compounds indicates that plants have a role in the nitrogen cycle. Nitrogen enters the plant with the water that the plant takes in through its roots. Nitrogen leaves the plant when it dies and decays or when parts such as leaves, twigs, or fruit drop from the plant. Plants also participate in the natural cycles for calcium, sulfur, phosphorus, and so forth. Each element that passes through the natural cycles spends some time in the soil, in water, in plants, and in animals. Part of the natural cycle of some elements takes place in the air. Plants obtain mineral matter from the soil. The water they absorb through their roots is not pure but is actually a solution of minerals that were in the soil. Plants will grow well only if the soil contains sufficient mineral matter of the right kinds. You will investigate the presence of minerals in soil in Project 3-8.

PROJECT 3-8: How much mineral matter does soil contain?

You will need: 100 grams of soil, a sensitive balance with a capacity of about 200 grams, masses for the balance, distilled water, a 500-milliliter graduated cylinder, a filter funnel, a filter flask, a filter pump, rubber tubing (to connect filter flask to filter pump), a stirring rod, a 1000-milliliter beaker, an evaporating dish with a capacity of about 400 milliliters, an electric hotplate with variable control, crucible tongs, an asbestos pad, a notebook, and a pen.

1. Measure out approximately 100 grams of soil on the sensitive balance. Record the mass of the soil. Put it into a 1000-milliliter beaker.
2. Add 200 milliliters of distilled water to the soil in the beaker.
3. Stir the soil and water well; then filter it as shown in Fig. 3-15.
4. Using the sensitive balance, measure and record the mass of the evaporating dish.
5. Pour the filtrate from Step 3 into the evaporating dish.
6. Put the filter paper and soil back into the beaker. Add 200 milliliters of distilled water to the filter paper and soil in the beaker; then repeat Steps 3 and 5.
7. Heat the filtrate to dryness on the electric hotplate. When almost all the water has evaporated reduce the temperature of the hotplate to avoid spattering.
8. When no water remains, turn off the hotplate. Use tongs to remove the evaporating dish from the hotplate. Place the evaporating dish on the asbestos pad and allow it to cool.
9. After the evaporating dish has cooled, measure and record the mass of the evaporating dish plus contents. Notice the grayish-white, ashy appearance of the contents. This indicates that they consist of mineral substances.
10. Subtract the mass of the evaporating dish from the mass of the evaporating dish plus contents to determine the mass of the contents.
11. Calculate the percentage of soluble mineral substances that the soil contained.

For Further Investigation

1. Perform Project 3-8 using several different types of soil (clay, loam, sandy). Compare the

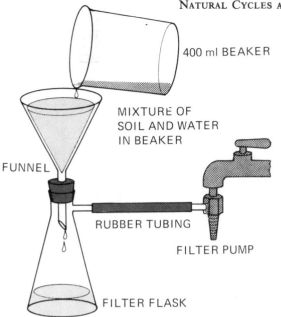

Fig. 3-15. Setup to filter a soil and water mixture.

percentages of soluble mineral substances in the different types of soils.

2. Perform Project 3-8 using a soil from a field or lawn that has been heavily fertilized.

3. Perform a project similar to Project 3-8 using water from ponds and streams instead of soil.

Think About This Problem

A salty crust often forms on the surface of soil that is regularly irrigated.

How might this problem affect you?
What causes the problem?
How could you help solve it?

Plants do not grow well on soil that contains too little mineral matter. For optimum plant growth, there must be a good deal of nitrogen, phosphorus, and potassium. Other elements, known as trace elements, are equally important but much smaller percentages of these elements are required.

To supply adequate amounts of the necessary mineral elements, farmers and other plant growers add fertilizer to the soil. Since inorganic fertilizer contains the mineral substances in easily soluble forms, much of these substances may be "leached" (washed) out of the soil during rainstorms. The dissolved minerals often flow into streams, ponds, and lakes and stimulate rapid growth of algae in the ponds and lakes. This condition, known as *eutrophication*, spoils those bodies of water for swimming, fishing, and even boating. Much more serious problems arise when dissolved fertilizer gets into the drinking water supply. For example, nitrates in the drinking water can cause cancer. Since interference with the natural cycles causes such harmful and dangerous results, a thorough understanding of the cycles is vitally necessary.

Humans have the ability to change some of the conditions in their environment. While these changes are often beneficial, they occasionally result in such harmful conditions as eutrophication of lakes and the production of cancer. Unlike humans, most organisms cannot change their environment at will, although they do have involuntary effects on their environment. The organisms can live only where the environment supplies what they require. Each species of organism has its own *niche*—or its way of using what its environment provides—and its ways of affecting its environment. Since organisms belonging to the same species need the same things from their environments, they compete with each other for what they require. To minimize the competition, some species of animals establish *territories* that they defend against all other members of their own species except for their mates and offspring. *Territoriality* is almost always successful against individuals of the same species. It is not an attempt to exclude animals that belong to different species. Under natural conditions, only one species of organism fills a particular niche in any one geographic area. Organisms belonging to different species do not compete with each other for food and other needs as seriously as organisms of the same species. There may, however, be violent and deadly conflict between organisms of different species if, for example, one species is a predator and the other species is its prey.

Every species of animal, except for the top carnivores (meat eaters), has at least one species of predator that may kill and eat it. The predator helps control the population of its prey species. In turn, as the number of prey animals decreases, the population of the predator species also decreases because less food is available to it. This mutual control of the populations of predator and prey is known as the *Balance of Nature*. In Chapter 4 you will investigate niches, territoriality, and the Balance of Nature.

CHAPTER 3. NATURAL CYCLES AND THE TRANSFER OF ENERGY

1. What is transpiration?

2. What happens to plants if their transpiration rate is greater than the rate at which they absorb water?

3. How is the rate of transpiration reduced naturally in some plants?

4. Why is it cooler under a leafy tree than under an awning when both are in the sunlight?

5. What are the two raw materials required for photosynthesis?

6. What are two products of photosynthesis?

7. What is a gaseous waste product of respiration?

8. Why may people help relieve the worldwide food shortage by eating vegetarian meals?

9. What is an herbivore?

10. About what percentage of an herbivore's food becomes additional mass of the herbivore?

11. Which nutrients contain only carbon, hydrogen, and oxygen?

12. Which nutrients contain mineral elements?

13. How does a plant obtain the mineral compounds that it needs for growth?

14. How may a lake be harmed if the nearby fields are heavily fertilized?

15. What are trace elements?

Chapter 4

NICHES, TERRITORIALITY, AND THE BALANCE OF NATURE

Look around you. All about are places where plants and animals may live. Out of doors, even in a crowded city, you will see an abundance of living things; grass, weeds, trees, birds, insects, and other forms of life are numerous. Some organisms can find ways of living even in the harshest environment. A small weed may force its way through a crack in a sidewalk. Alongside it, ants may scurry in and out of the entrance to their underground home.

To create a favorable environment in our homes, we try to exclude all organisms except other people, our pets, and our houseplants. However, the thriving businesses conducted by exterminators and insecticide manufacturers show that we are not always successful in getting rid of organisms such as cockroaches, bugs, and rodents, which we consider undersirable. Even in the cleanest buildings, a variety of organisms can manage to live.

In any environment, each species of plant or animal lives in its own special way, using what it needs in its environment and, in turn, having certain effects on the environment. The organism's way of life is its *niche,* which is frequently described as the organism's profession. Although many different species of animals may live in the same habitat, each lives differently from the others. For example, on a lawn, earthworms live mainly in underground tunnels, feeding on the organic matter in the soil. Ants also have underground tunnels but they may eat seeds, insects, fungus, or a variety of other food, depending on the species of ant. Robins forage mainly for earthworms and insects on the lawn, while a species of bird called the golden-shafted flicker eats almost nothing but ants. An animal's food preferences comprise an important facet of its niche. Some of these food preferences may harm us. Others might be helpful. In Project 4-1 you will investigate and identify at least one insect whose niche includes a preference for eating fruit. This insect might well spoil a fruit crop, cause a fruit shortage on the market, and help drive up food costs. We will also identify at least one insect that prefers to eat meat. This insect might be helpful because it can decrease food decay and the spread of harmful bacteria.

PROJECT 4-1: Some food preferences of insects.

You will need: two glass jars, two 20-centimeter square pieces of ½-inch hardware cloth (wire mesh, with wires ½ inch apart), a pair of pliers, a pair of work gloves, a banana, a piece of meat with a mass of about 30 grams, two 20-centimeter square pieces of nylon stocking, two rubber bands, a notebook, and a pen.

NOTE: This project is best performed during late spring, the summer, or early autumn.

1. Place the pieces of hardware cloth on the tops of the glass jars.
2. Put gloves on to protect your hands. Then, with the pliers, bend down the edges of the hardware cloth, forming covers for the jars (Fig. 4-1).
3. Remove the hardware cloth covers from the jars. Put a piece of banana into one jar and a piece of meat into the other jar. Replace the hardware cloth covers on the jars (Fig. 4-2).
4. Set the jars outdoors.
5. Observe the jars daily, noting and recording the kinds of adult insects each food attracts.
6. During this period, you may observe larvae which developed from eggs laid on the food by adult insects. If so, replace the hardware cloth with the squares of nylon stocking to prevent the escape of the larvae or the adults that will develop later. Fasten the nylon over the mouth of each jar with a rubber band; then bring the jars indoors and set them in a room at room temperature. Allow the larvae to pupate and then to become adults. This should take from 10 to 15 days.
7. Observe the adults that appear in each jar. Compare their appearance to that of the adults that were attracted to the food in each jar in Step 5 (Fig. 4-3).
8. Explain your results.

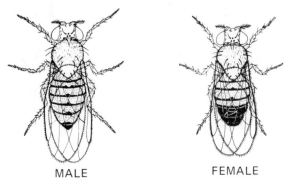

Fig. 4-3. Adult drosophila (fruit flies). (Courtesy Carolina Biological Supply Co.)

For Further Investigation

1. Repeat Project 4-1 using other kinds of fruit instead of bananas. Investigate the effectiveness of different kinds of fruit in attracting fruit flies.

2. Perform projects similar to Project 4-1 using liver, fish, and honey in individual jars in place of meat and bananas.

3. Watch for larvae feeding on garbage in outdoor garbage cans as well as on other foods and in other places. Put some of the food, along with some larvae, into a jar. Cover the jar with nylon stocking material and raise the larvae to adulthood. Observe the kinds of adults that appear.

Think About This Problem

Many houses are severely damaged by termites (Fig. 4-4).

How might this problem affect you?
What causes the problem?
How could you help solve it?

Fig. 4-1. Bending hardware cloth to fit over jars.

Fig. 4-2. Covered jars containing food.

Some insects, such as cockroaches, will eat practically anything. They are almost completely unspecialized in their choice of food. In contrast, the insects you investigated in Project 4-1 show some degree of specialization in their food preferences. Fruit flies, for example, feed and breed on fruit but not on meat. Many different kinds of

Fig. 4-4. Termites damaging a wooden beam.   (*Courtesy American Museum of Natural History*)

fruit, however, will serve as food for the fruit flies. Cabbage butterflies have much more specialized food preferences. This is what you will investigate in Project 4-2.

PROJECT 4-2: Food preferences of the cabbage butterfly caterpillar.

You will need: cabbage seedlings, gardening tools, a thermometer, an old nylon stocking, a pair of scissors, thin (#20) wire, an insect cage, a dish, a cellulose sponge, a metric rule, some leaves of cabbage, carrots, broccoli, cauliflower, lettuce, spinach, and wild mustard, a notebook, and a pen.

1. Prepare a small garden patch. A window box will do if it is kept outdoors. Plant cabbage seedlings in the garden when the soil temperature is at least 18° Celsius and at least 30 days before the first frost.
2. Water, weed, and fertilize (not too heavily) the cabbage plants to encourage them to grow vigorously.
3. Watch for small holes in the cabbage leaves. If you find them, look for small, green caterpillars chewing at the edges of the holes. These caterpillars are usually found on the center leaves. They are most likely to be cabbage butterfly caterpillars (Fig. 4-5).
4. Cut a 20-centimeter tubular piece from an old nylon stocking. Close one end of the tube by

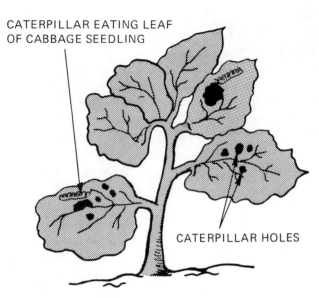

Fig. 4-5. *Caterpillars chewing holes in cabbage leaves.*

twisting a piece of wire around it. Slip the open end of the tube over the part of a cabbage plant that has some caterpillars feeding on it. Close the lower end *loosely* around the cabbage stem by twisting another piece of wire around it (Fig. 4-6). This procedure, known as *sleeving*, protects the caterpillars against predators and also prevents the caterpillars from leaving a particular plant.

5. When the caterpillars are about 1 centimeter long, remove a few from the cabbage plants and place them in an insect cage indoors. Now put a leaf of each of the following kinds of plants into the cage: cabbage, carrot, broccoli, cauliflower, lettuce, spinach, and wild mustard. Note and record which kinds of leaves the caterpillars eat.

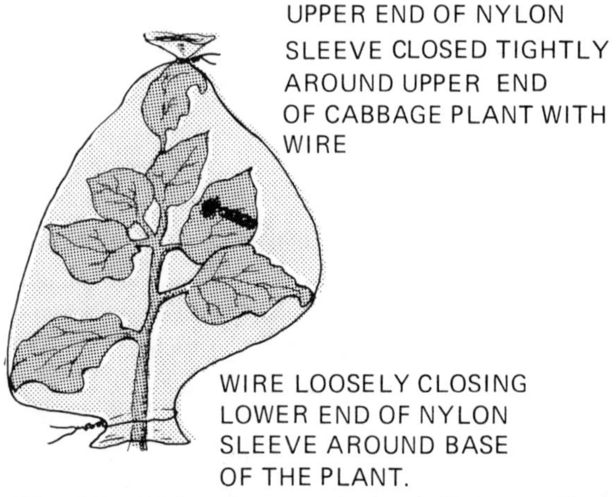

6. Explain your results.

For Further Investigation

1. All spiders are carnivores, feeding mainly on insects. Yet each has its own niche, with different types of webs (or none at all) and different ways of catching their prey. Investigate such differences among the house spider, garden spider, wolf spider, crab spider, and others.

2. Houseflies and cockroaches both spend most of their time in buildings and both feed on many of the same types of foods yet their niches are different. Describe some of the differences between their niches.

3. Based on what you know about the food choices of the cockroach, explain why it can exist in large numbers in buildings.

Think About This Problem

If the same kind of plant is grown year after year in the same location, it tends to be damaged more each year by insects.

How might this problem affect you?
What causes the problem?
How could you help solve it?

Cabbage butterfly caterpillars eat only plants of the mustard family: cabbage, broccoli, cauliflower, and wild mustard. They will not eat plants belonging to any other family. Other insects are even more specialized, feeding on only one species of plant or animal and only in a certain way. Gall insects are excellent examples of such specialized animals. Some of the goldenrods that grow in many fields and empty lots have galls, which are swellings composed of plant tissue. Galls develop after an adult insect lays an egg inside the goldenrod stem. The larva that hatches from the egg lives inside the gall, feeding on the plant tissue. To learn more about its specialization, we will investigate the niche of a gall insect in Project 4-3.

Fig. 4-6. *Nylon sleeve enclosing cabbage seedling.*

PROJECT 4-3: Checking on specialization in gall insects.

You will need: some old nylon stockings, a pair of scissors, thin (#20) wire, a notebook, and a pen.

1. Find ten goldenrods with stem galls. Figure 4-7 shows what goldenrod stem galls looks like.
2. Cut ten 20-centimeter tubular pieces from the stockings. These cut pieces are known as sleeves (Fig. 4-8).
3. Slip a stocking sleeve over each of the ten goldenrods so that the gall is in the middle of the tube. Fasten the top and bottom of the tube around the stem of the goldenrod with pieces of wire.
4. Inspect the goldenrod plants at least twice each week until you notice insects inside the stocking sleeves. Compare the insects that are in the ten tubes. Are they all alike or are they different?
5. Explain your results.

For Further Investigation

1. Perform projects similar to Project 4-3 using different kinds of galls, such as the oak leaf gall.

2. The larva of certain insects curls a leaf, then pupates inside the leaf, and finally emerges as an adult. Locate some leaves that are curled in this way. Sleeve the leaves with stocking tubes, wired to the stems. Observe the adults that emerge from the curled leaves.

Think About This Problem

Under certain conditions, insects (such as the European elm bark beetle) that feed exclusively on one plant species may almost exterminate that species of plant.

How might this problem affect you?
What causes the problem?
How could you help solve it?

Gall insects use their environments very effectively. For example, the adult female goldenrod gall insect lays an egg inside the stem of a goldenrod plant. This stimulates the stem to swell around the egg. When the larva emerges from the egg, it feeds on the tissues within the enlarged portion of the stem. As it grows, the larva is protected from birds and other predators because it is surrounded by the tough stem (Fig. 4-9). Although the gall insect uses its environment to its advantage, it also affects its environment by making the goldenrod stem swell. The goldenrod stem gall

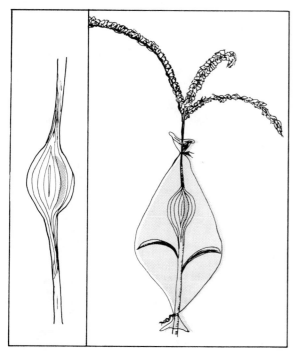

Fig. 4-7. Goldenrod stem gall.

Fig. 4-8. Nylon sleeve closing top and bottom of goldenrod stem gall.

Fig. 4-9. Longitudinal section drawing showing larva curled up inside goldenrod gall.

insect has a harmful effect on the goldenrod. Some other insects have beneficial—even essential—effects on their environments. Such effects are parts of the organism's niche, whether the effects are harmful or beneficial. In Project 4-4 you will investigate an essential, beneficial environmental effect in an insect's niche.

PROJECT 4-4: The niche of the bee.

You will need: an old nylon stocking, a pair of scissors, thin (#20) wire, a magnifying glass, a notebook, and a pen.

NOTE: This project is best performed during the spring, when dandelion blossoms are most abundant.

1. Cut a 10-centimeter tubular piece from an old nylon stocking, forming a sleeve. Close one end of the sleeve by twisting a piece of wire around it.

2. Slip the open end of the sleeve over an unopened dandelion bud. Close the lower end *loosely* around the dandelion stem by twisting another piece of wire around it (Fig. 4-10).

3. Within a few days, the sleeved bud will become a flower. Watch for bees or other insects that may walk on the nylon sleeve or visit other (unsleeved) dandelion flowers (Fig. 4-11).

4. After the flower in the nylon sleeve has lost its petals, examine it carefully with a strong magnifying glass to see whether seeds have developed. Compare the flower with another dandelion flower at a similar stage that has not been sleeved.

5. Explain your results.

Fig. 4-10. Nylon sleeve over dandelion plant. (Photo by Alan J. Federow)

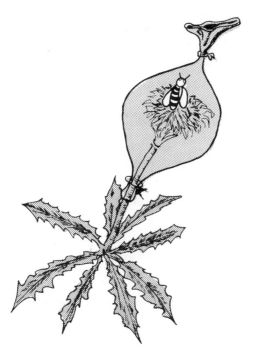

Fig. 4-11. Bee walking on outside of nylon sleeve enclosing dandelion flower.

For Further Investigation

1. Repeat Project 4-4, using other flowering plants (e.g., wild mustard, cherry, blackberry) instead of the dandelion.

2. Perform projects similar to Project 4-4 but make the sleeves of fiberglass insect screening instead of nylon stocking material. The insect screening will not prevent wind pollination, although it will prevent insects from pollinating the plants. As subjects, use some plants with showy flowers (dandelion, wild mustard, cherry, blackberry, etc.) and other plants that have inconspicuous flowers which are generally green (maple, grass, corn, plantain, etc.). Determine which flowers are insect-pollinated and which are wind-pollinated.

3. Sleeve unopened flower buds with nylon tubes. Keep them sleeved each day from one-half hour before sunrise to one-half hour after sunset. Include among your subjects some flowers that open late in the day (e.g., four-o'clocks). Determine which flowers are pollinated at night.

4. Pull the parachutelike fruits from a ripe dandelion head. Count the number of fruits that have black seeds at the bottom and the number that have white, undeveloped seeds. Express the ratio of black seeds to white, undeveloped seeds as a percentage. This is the same as the percentage of florets (little flowers) on the dandelion head that were pollinated. It is a measure of how effectively bees pollinate the flowers. Find out whether this percentage varies significantly in different localities.

Think About This Problem

The number of beekeepers in the United States has been decreasing.

How might this problem affect you?
What causes the problem?
How could you help solve it?

Insects perform an absolutely essential service by pollinating plants. Although butterflies, moths, flies, and other kinds of insects pollinate some plants, the honeybee is by far the most important and effective pollinator. Where there are few honeybees, there are poor crops of apples, peaches, and many other kinds of fruits and vegetables. Farmers and orchard owners often rent hives of honeybees from beekeepers. After the flowers have been pollinated on a farm or orchard, the beekeeper moves the hives to a different location, where the bees can pollinate other kinds of flowers.

Bees affect their environments by pollinating flowers and making it possible for the plants to reproduce sexually. After pollination, ovules in the female parts of the flowers are fertilized and then develop into seeds. Bees help maintain or increase their own food supply by pollinating flowers. The seeds that develop may grow into plants whose flowers will provide the bees with additional nectar and pollen.

Many other kinds of organisms change their environments in ways that are harmful to the organisms themselves. These same changes improve the environments for some other organisms, which, in turn can thrive and replace the original organisms. For example, the cattail plants that grow in a pond may gradually help fill in the entire pond. This prevents other cattails from growing but allows the growth of alders, willows, and other plants. The series of changes in the kinds of organisms that successively inhabit the changing environment is termed "succession." When we understand the kinds of changes that can be expected to take place during the succession in each type of environment, we can prepare for these changes. Thus we may be able to prevent undesirable changes or utilize the desirable changes to our best advantage. In Project 4-5 you will investigate one form of succession, the way yeast living in a molasses solution prepares the environment for certain bacteria.

PROJECT 4-5: Succession in a molasses solution.

You will need: a bottle of molasses (without preservatives), distilled water, a package of yeast, a 1000-milliliter Erlenmeyer flask, a 500-milliliter graduated cylinder, a stirring rod, a microscope, a medicine dropper, a culture dish, microscope slides, cover slips, lens cleaning tissue, methylene blue solution, blue litmus paper, red litmus paper, a notebook, a pen, and a pencil.

1. Measure 10 milliliters of distilled water in the graduated cylinder and pour it into the culture dish.

2. Add a pinch of yeast to the water in the culture dish. Stir the water and yeast well with the stirring rod.

3. Prepare a slide of the yeast and water, staining it with methylene blue. (See Fig. 4.12.)

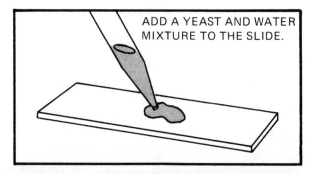
ADD A YEAST AND WATER MIXTURE TO THE SLIDE.

ADD METHYLENE BLUE SOLUTION TO THE WATER AND YEAST MIXTURE.

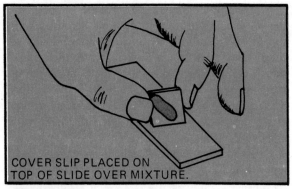
COVER SLIP PLACED ON TOP OF SLIDE OVER MIXTURE.

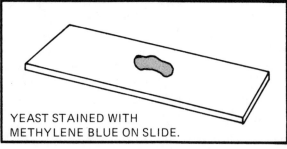
YEAST STAINED WITH METHYLENE BLUE ON SLIDE.

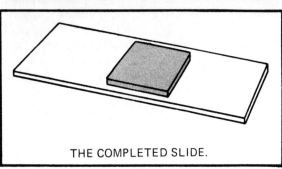
THE COMPLETED SLIDE.

Fig. 4-12. Preparing a slide of yeast.

4. Examine the slide under both the low and high powers of the microscope. Make drawings of some yeast cells in your notebook (Fig. 4-13).

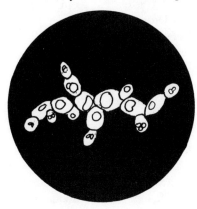

Fig. 4-13. Drawing yeast cells as seen under microscope.

5. Measure 500 milliliters of distilled water in the graduated cylinder, and pour it into the Erlenmeyer flask.

6. Add the remainder of the package of yeast to the water in the flask and stir the water and yeast well with the stirring rod.

7. Test the yeast-in-water culture with blue litmus paper and with red litmus paper to find out if the culture is acidic, basic, or neutral. Record your finding in your notebook.

8. Put the culture in a warm (20° Celsius to 35° Celsius), dark place.

9. Examine the culture once each day as follows:
    a. Smell the culture, noting the presence of an odor like vinegar or like alcohol.
    b. Test the culture with litmus paper as in Step 7.
    c. Prepare a microscope slide of the culture, stained with methylene blue, and examine it under the microscope. Note any increase or decrease in the number of yeast cells. Note the presence and relative numbers of bacteria, which will probably be rod-shaped; they may, however, be spherical.
    d. Record your findings in your notebook.

10. Write a description of the changes that take place in the culture.

11. Explain your results.

For Further Investigation

1. Make a hay infusion by boiling a few pieces of dry grass stem in distilled water. Innoculate the infusion with a few drops of water from a pond. Examine the infusion once each day with a microscope. Note the succession of organisms that occur in the infusion.

2. Locate an empty lot or field where the plants have not been cut for 2 or 3 years. Locate an empty lot or field where the plants have not been cut for about 10 years. Note the differences among the species of plants that are growing in the two locations.

Think About This Problem

Some lumber companies are demanding the right to cut down giant redwood trees, which are the climax stage of a succession in a limited area in the United States (Fig. 4-14).

Fig. 4-14. Lumbering operations in a giant redwood forest.

(Courtesy American Museum of Natural History)

How might this problem affect you?
What causes the problem?
How could you help solve it?

The bacteria that replace yeast in a molasses solution, as in Project 4-5, do not attack or damage the yeast cells. The bacteria simply take advantage of the fact that the yeast has changed some of the sugar (in the molasses) into alcohol. The alcohol is food for this kind of bacteria, which obtain energy by changing alcohol to vinegar (acetic acid). As the yeast changes sugar into alcohol, it not only depletes its food supply but it also inhibits its own growth because the yeast cannot grow and reproduce where the alcohol concentration is too high. Thus, in improving the environment for the bacteria, the yeast ruins its own environment. That is why the bacteria replace the yeast.

Year after year, as plants grow in a field, taller plants overshadow shorter ones and prevent the shorter plants from growing. Typically, in this changing process, "weeds" and grasses are followed by shrubs, which are followed, in turn, by trees. "Pioneer" trees (birch, sumac, poplar, etc.), whose seedlings can grow only in ample light, are followed by "climax" trees, whose seedlings can grow only in the shade.

All animals (except for top carnivores) have predators that kill and eat them. The interactions between predator and prey result in the control of the populations of each species, which is known as the Balance of Nature. For example, the Canada lynx and the snowshoe rabbit help control each other's populations. In a year when snowshoe rabbits are abundant, the lynx which prey on them have an ample supply of food and can survive and reproduce. The population of lynx is greater during the next year. Since they require more food, they kill more rabbits, reducing the population to the point where there are not enough rabbits to supply food for the lynx. Consequently, some lynx starve, some do not reproduce, and the population of lynx decreases to a low level. Now the scarcity of predators allows the population of rabbits to increase rapidly and the cycle begins once again. You will observe how the Balance of Nature operates in the case of two other species in Project 4-6.

PROJECT 4-6: Paramecium and Didinium—Balance of Nature in action.

You will need: distilled water, two dried lima beans, a sheet of newspaper, a hammer, a paramecium culture, a didinium culture, two 400-milliliter beakers, three pipettes (medicine droppers), a binocular microscope, a standard microscope, microscope slides, cover slips, methylene blue solution, a notebook, a pen, a pencil, and graph paper.

1. Fold the newspaper several times and put it on a surface that will not be damaged by hammering. Put two dried lima beans inside the folded sheet of newspaper. Hammer the beans to break them into pieces within the newspaper.

2. Pour distilled water into one beaker until it is about half full. Put the broken beans into the water and put the beaker in a warm (20° Celsius to 35° Celsius), dark place. In about 2 days, bacteria should grow in the water-and-bean mixture, producing what is known as a "bean culture."

3. Make a slide with 2 drops of paramecium culture, staining it with methylene blue.

4. Examine the slide under the standard microscope at low and high power. Draw a paramecium in your notebook (Fig. 4-15).

5. Repeat Steps 3 and 4 using didinium culture (Fig. 4-16).

6. Pour distilled water into the second beaker until it is about half full. Using an individual clean pipette for each culture, add 20 drops of bean culture and 10 drops of paramecium culture to the water in the beaker (Fig. 4-17). Rinse the pipettes after each use.

7. Put the beaker (prepared in Step 6) in a warm, dark place. Once a day, using a clean

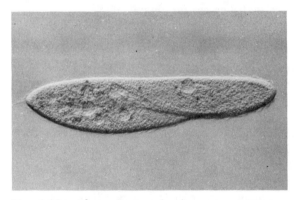

*Fig. 4-15. Photomicrograph of a paramecium.
(Courtesy Carolina Biological Co.)*

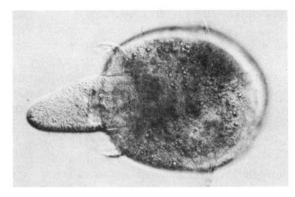

Fig. 4-16. Photomicrograph of didinium eating a paramecium.   (Courtesy Carolina Biological Co.)

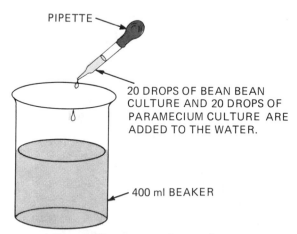

Fig. 4-17. Adding paramecium culture to water-and-bean culture.

pipette, put 5 drops of water from this beaker into a watchglass and use the binocular microscope to count the number of paramecia present in the water. Calculate and record the number of paramecia per drop of water.

8. Each day during the duration of the project, add 20 drops of bean culture to the water in the beaker.

9. When there are at least five paramecia per drop of water in the beaker, add 20 drops of didinium culture to the water in the beaker.

10. As before, keep the beaker in a warm, dark place and add bean culture each day. Also, continue to count and record the number of paramecia per drop of water. At the same time, count and record the number of didinia per drop of water. Discontinue the project when the number of paramecia have decreased below one per drop.

11. Graph the number of paramecia per drop of water against time in days. (See Fig. 4-18.)

12. Graph the number of didinia per drop of water against time in days (Fig. 4-19).

13. Explain your results.

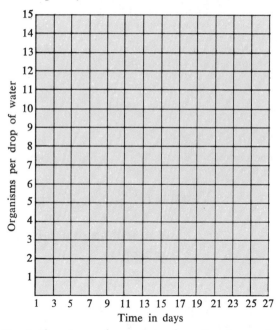

Fig. 4-18. A sample graph to show the changes in the population of paramecia.

For Further Investigation

1. Watch a rosebush in springtime to observe when aphids (plant lice) first infest it. Each day, count and record the number of plant lice on one twig. Put some ladybird beetles on the plant. Continue counting and recording the number of aphids and the number of ladybird beetles on the twig. Graph your findings.

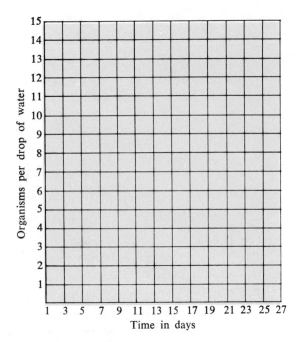

Fig. 4-19. *A sample graph to show the changes in the population of the didinia.*

2. Do reading research to find out more about the relationship between rabbits and lynx.

Think About This Problem

The number of deer in a locality often becomes so large that they cannot obtain sufficient food during the winter.

How might this problem affect you?
What causes the problem?
How could you help solve it?

Project 4-6 demonstrated that didinia introduced into a paramecium culture rapidly reduces the number of paramecia. When paramecia become scarce, however, the didinia die off. Predators must have an ample, dependable supply of prey animals in order to maintain their own numbers. One way the predators can insure a sufficient food supply is to claim an adequate hunting area and defend it against other members of their own species. Like many other animals, the male red fox marks its territory with urine to warn away other adult male red foxes. The only members of its own species the adult male tolerates within its territory are its mate and their offspring. Certain predators, such as wolves, hunt in packs. In such cases, the entire pack claims a territory. Some herbivores, including beavers, claim and mark territories to insure a dependable food supply. Territoriality, the claiming and defense of a territory, is an important factor in the survival of many kinds of animals. In Project 4-7 you will investigate territoriality in a bird species.

PROJECT 4-7: How the robin claims and defends its territory.

You will need: a pair of binoculars, a notebook, a pencil, and a pen.

NOTE: This project can be done only in areas where robins nest and breed (Fig. 4-20). In other areas different species of birds will be suitable subjects for this project.

1. Beginning when robins first appear in sizable numbers, record the date and number of robins per group each time any robins are observed.
2. Record the date and time of day when you hear any robins singing.
3. After a few weeks, you will observe that no group of robins consists of more than two birds. This indicates that mating pairs have formed and territories have been chosen. Begin looking for

Fig. 4-20. *Robins on their nest.*

robin nests at this time. You can locate a nest by watching the birds and noticing the location they repeatedly fly back to after foraging for food. A bird with nesting material (grass, moss, etc.) in its beak will fly to the nest it is building. Use the binoculars to verify that the nest is where you believe it to be. Make a note of its location.

4. After locating one robin's nest, look for other robins in the surrounding areas. When you find them, locate their nests and record the locations.

5. Measure the approximate distances between pairs of robins' nests. This may be done by a number of methods, including pacing off the distance between nests or by marking the location of each nest on a map of the locality and using a scale to estimate the distance.

6. Explain why there is a certain minimum distance between pairs of nests.

7. Describe any evidence that the robins' singing occurs only when they are courting or are paired and raising their young.

For Further Investigation

1. Locate woodchuck holes in a field. Measure the distances between pairs of holes.

2. Observe ways in which male cats and dogs mark and defend their territories against other males of the same species.

Think About This Problem

The nesting territories of seven pairs of sandhill cranes are located directly in the path of the proposed route of Highway I-10 in Mississippi. Only 40 of these birds still exist under natural conditions.

How might this problem affect you?
What causes the problem?
How could you help solve it?

We may think of the robin's song as something beautiful that probably expresses the bird's joy and enthusiasm. This is not so. Actually, the robin's song is more of a war chant and threat. Other male robins rarely approach a male robin that is singing to advertise his territory. Only his mate and offspring are attracted rather than driven away by his song. If an intruder enters an occupied territory, a fight ensues. As far as we know, the owner of the territory is invariably the victor.

It is difficult—if not impossible—to determine how members of some other species mark their territories and warn other members of their species to stay away. These species must accomplish these things in some way unknown to us because they definitely have territoriality. The yellow jacket hornet is one such species. You will investigate evidences of its territoriality in Project 4-8.

PROJECT 4-8: Territoriality among yellow jacket hornets.

You will need: tuna fish, a dish, a topographic map of the area (see Project A-1 in the Appendix: How to obtain topographic maps), a magnetic compass, a ruler, tracing paper, masking tape, a drawing board mounted on a tripod, an alidade (see Project A-2 in the Appendix: Making a plane table survey), a notebook, a pen, and a pencil.

NOTE: Yellow jacket hornets are quite gentle and do not sting unless they are seriously provoked. If you remain aware of where the yellow jackets are and avoid making any sudden moves that might startle them, you will not be stung. Of course, people with severe allergy to insect stings should be especially cautious.

1. On the topographic map that represents the area you are investigating, draw several parallel lines, about 2 centimeters apart, pointing toward magnetic north.

2. Use the following method to find the point on

the map that represents your position in the field.

a. Set up the drawing board and tripod at your position in the field. Attach the map to the drawing board with masking tape.

b. Place the compass on the map. Turn the drawing board and map until the compass needle is parallel to the lines you drew pointing to magnetic north (Fig. 4-21).

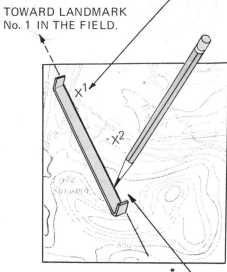

Fig. 4-22. *Fixing your position A on the topographic map.*

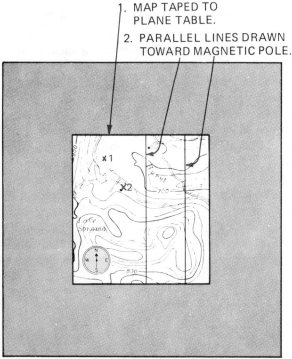

Fig. 4-21. *Indicating magnetic north on the topographic map.*

c. Place an alidade on the map so that its working edge is on the point representing a landmark shown on the map. Sight along the alidade toward the actual landmark, turning the alidade until it points toward the landmark. Draw a line in that direction through the point representing the landmark on the map (Fig. 4-22).

d. Repeat Step 2c with a second landmark. The point where the two lines intersect indicates your position on the map. Extend the lines if they do not intersect as originally drawn. Draw a circle around the point and label it as your position A (Fig. 4-23).

3. Put some tuna fish in a dish. Place it on the ground at your location in the field.

4. Watch for yellow jacket hornets (Fig. 4-24) which may arrive to feed on the tuna fish. Notice the direction in which the hornets fly away. They are probably flying directly back to their nest. Place the alidade on the map so that its working edge is on the point representing your position. Sight in the direction the hornets fly away. Draw a line on the map along the edge of the alidade to represent that direction (Fig. 4-25).

5. Move to another location, about 100 meters away (Fig. 4-26). Repeat Step 2 to find the point on the map that represents your new location B (Fig. 4-27).

6. Put the dish of tuna down at your new location. Repeat Step 4. The probable location of the hornets' nest is represented on the map by the point where the lines from your two field locations intersect (Fig. 4-28).

7. Move several hundred meters from your

Niches, Territoriality, and the Balance of Nature

1. ALIDADE WITH ONE EDGE ON LANDMARK No. 2 ON MAP SIGHTED TOWARD LANDMARK No. 2 IN FIELD.

TOWARD LANDMARK No. 2 IN THE FIELD

2. LINE DRAWN ALONG EDGE OF ALIDADE THROUGH LANDMARK No. 2 ON MAP.

3. POINT ON MAP WHERE TWO LINES INTERSECT INDICATES YOUR POSITION A IN THE FIELD.

Fig. 4-23. Fixing your position A on the topographic map (continued).

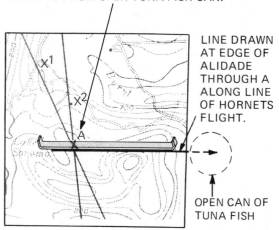

ALIDADE WITH ONE EDGE PASSING THROUGH A ON MAP, SIGHTED ALONG LINE OF FLIGHT OF HORNETS FROM OPEN TUNA FISH CAN.

LINE DRAWN AT EDGE OF ALIDADE THROUGH A ALONG LINE OF HORNETS FLIGHT.

OPEN CAN OF TUNA FISH

Fig. 4-25. Sighting direction in which hornets fly from open can of tuna fish at position A.

original location. Repeat Steps 2, 3, 4, 5, and 6 to locate additional hornets' nests.

8. Use the ruler and the scale printed on the topographic map to determine the distances between pairs of hornets' nests.

9. Explain why the results of your investigation are evidence of territoriality among yellow jacket hornets.

For Further Investigation

1. Locate honeybee hives by observing the direction in which the bees fly after gathering nectar

Fig. 4-24. Yellow jacket hornet (left). Yellow jacket hornet's nest in the ground (right).
(Courtesy American Museum of Natural History)

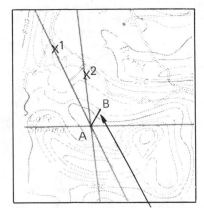

Fig. 4-26. *Moving to your second position B in the field.*

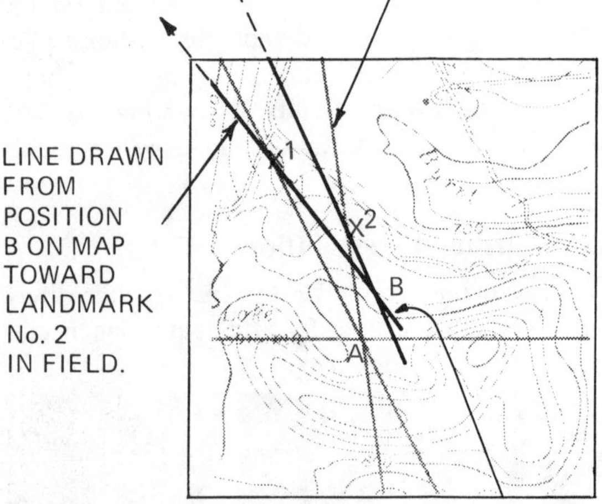

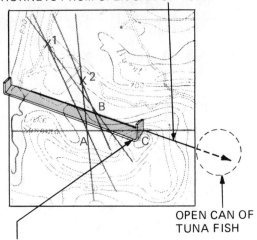

Fig. 4-28 *Fixing the hornets' nest location on the topographic map.*

Fig. 4-27. *Locating your position B on the topographic map.*

from flowers. The bee hives will generally be found in hollow beetrees. Remember that bee trees are sometimes hives that belong to beekeepers. Use methods similar to those outlined in Project 4-8.

2. Photograph birds sitting on telephone wires or other horizontal objects. Measure the distances between pairs of birds. Explain why these distances are quite uniform.

Think About This Problem

Draining swamps destroys many of the breeding territories of frogs (Fig. 4-29).

How might this problem affect you?
What causes the problem?
How could you help solve it?

Members of any species that has territoriality will not permit most other members of the same species to enter their territory. They are unconcerned, however, by the presence of different species of animals all around them. Each species of animal has its own niche that is different from all other niches in the same habitat. Therefore, the various species do not interfere with each other, except, of course, if one species preys on, or parasitizes, another.

There are many possible niches in any one habitat and a correspondingly large number of species of organisms, both animal and plant, which fill these niches. All these species constitute the community.

As more of the land is developed for building

Fig. 4-29. Frogs in a swamp. Find the egg masses and the young tadpoles.
(Courtesy American Museum of Natural History)

construction and agriculture, fewer niches remain in the community and there is a corresponding decrease in the number of species in that community. As we diminish the population of the various species we do harm to ourselves as well as the other organisms. A clear understanding of communities and how they affect each other will help make this a more livable world for all species, including humans. In Chapter 5 you will investigate some communities and their interrelationships.

CHAPTER 4. NICHES, TERRITORIALITY, AND THE BALANCE OF NATURE

1. What is an ecological niche?

2. Why can many different species of organisms live in the same habitat?

3. How do fruit flies, cockroaches, and cabbage butterfly larvae differ in the degree of specialization of their food preferences?

4. What causes a goldenrod gall?

5. Why do certain insects curl the leaves of some plants?

6. How can you tell how effectively bees have pollinated dandelion flowers?

7. Why do farmers sometimes rent hives of honeybees from beekeepers?

8. How do honeybees benefit when they pollinate flowers?

9. What is ecological succession?

10. Why can certain bacteria thrive in a molasses solution after yeast has been growing there for a while?

11. What are pioneer trees?

12. How do rabbits and lynx help control each other's populations?

13. Describe the territoriality of the red fox.

14. Why does a male robin sing?

15. What is an ecological community?

Chapter 5
COMMUNITIES

An empty lot or an old, uncultivated field that you pass every day and scarcely notice can be an endless source of entertainment and instruction. There may be hundreds of different sorts of plants and animals in it. As you observe such an area carefully, you may become aware of a vast number of problems you might investigate. What encourages the growth of the ragweed that causes the agonies of hayfever? Why are there many goldenrods in one part of the field and none at all in another part? How did the young birch trees begin growing in the field? Where can we find the most crickets? These are but a few of the many questions that might arise. As you attempt to answer such questions, you gain a better understanding of the communities that exist in different habitats. This improved understanding is highly important to you because you are a part of the community you are investigating. You live (at least temporarily) in the same habitat as the other organisms. What affects the rest of the community also affects you directly or indirectly.

There are so many species in most habitats that it is difficult to investigate more than one aspect of the community at a time. Each project in this chapter will direct your attention to just one characteristic of a community. Investigating these individual aspects can help you better understand the nature of the whole community. Armed with this understandng, you may learn how to manage various habitats in ways that are in your best interests as well as those of the other species that comprise these communities.

You will begin in Project 5-1 by investigating a habitat that is small in size but has a large number of different organisms in its community.

PROJECT 5-1: Which species make up the community in a rotten log?

You will need: five collecting jars, a magnifying glass, a large screwdriver, a woodsman's saw, a notebook, a pencil, a pen, and a metric rule.

1. Find a rotten log on the ground (Fig. 5-1).
2. Write a brief description of any organisms found on the top and side surfaces of the log. Estimate the number of each organism on the log and record the number in your notebook. (Rapid pencil sketches of the organisms will be helpful.)
3. Roll the log over. Follow the procedure in Step 2 for the organisms you find on the ground where the log originally lay (Fig. 5-2).
4. Follow the procedure in Step 2 for the organisms you find on the upturned surface of the log (which was originally on the bottom).
5. Break away part of the upper surface of the log with the screwdriver to a depth of about 1

COMMUNITIES

Fig. 5-1. A rotten log. What kinds of organisms can you see on this log?
(Courtesy American Museum of Natural History)

Fig. 5-2. A millipede and other organisms under a rotten log.   (Courtesy American Museum of Natural History)

centimeter. Follow the procedure in Step 2 for the organisms you have disclosed at this depth.

6. Break away part of the lower part of the lower surface of the log to a depth of about 1 centimeter. Follow the procedure in Step 2 for the organisms you have disclosed.

7. Break away rotten wood until you reach a layer of firm wood. Follow the procedures in Step 2 for the organisms you have disclosed.

8. Cut through the log with the woodsman's saw. Follow the procedures in Step 2 for the organisms you find within the solid part of the log.

9. Draw a diagram of a cross section of the log, showing the kinds of organisms you found in the different regions.

10. Explain why the different kinds of organisms live in the different regions of the log.

For Further Investigation

1. Perform a project similar to Project 5-1 but instead of drawing the organisms take extreme closeup photographs of them.

2. Perform projects similar to Project 5-1 using logs of the same species in different stages of decay. Compare the communities you find in each stage of decay. Explain how your results illustrate succession in a rotten log.

3. Perform projects similar to Project 5-1 using rotten logs of different species of trees.

4. Put a wooden board on the ground and allow it to stay there for several weeks. Compare the organisms you find under the board with those found under a rotten log.

5. Perform projects similar to Project 5-1 with one log in a shady forest and another log in a sunny, open field. Compare your findings in the two logs.

6. Establish a compost pile, use grass clippings, leaves, and table scraps. Compare the organisms found in the compost pile after about 1 month with those found in a rotten log.

Think About This Problem

In developed areas dead trees are removed instead of being permitted to lie on the ground.

How might this problem affect you?
What causes the problem?
How could you help solve it?

We have observed that a rotten log is a small, compact habitat in which many different species live because a large number of niches are possible there. It is interesting to observe how the number of possible niches and, therefore, the number of different species changes as the rotting process progresses. An initial increase in niches and species is followed by a decrease in both niches and species as the log approaches complete disintegration. The number of species of organisms surviving in a log under changing conditions depends on the number of possible niches as the conditions change.

Although a field is much larger and is a much different habitat than a log, the same general principles apply to both kinds of habitat. Both the rotten log and the field have communities that consist of large numbers of different species of organisms. Different parts of both habitats, the field and the log, may have different conditions that determine which species can live there. In Project 5-1 you probably found that the kinds of animals living in the interior of the log were different from those living on its surface. This was undoubtedly due to the different conditions on the exterior and in the interior of the log. In Project 5-2 you will investigate the kinds of plants that live in a field. If conditions are not the same in various parts of the field, you may expect to find different species of plants in each location. You will select two different parts of the field and compare the kinds of plants that live in each part. Your results will be expressed as an *Index of Similarity*, which tells how much alike the plants are in the two parts of the field. You will also express your results as an *Index of Dissimilarity*, which tells how different the plants are.

COMMUNITIES

PROJECT 5-2: How to determine similarity and dissimilarity among plants in different locations.

You will need: a quadrat frame (see Project A-5 in the Appendix: Making a quadrat frame), a notebook, and a pen.

1. Select a field where you wish to investigate similarity and dissimilarity among plants.
2. Choose any location in the field. Flip the quadrat frame backward over your head.
3. See Project A-6 in the Appendix: Differentiating plants. Using the methods given in Project A-6, differentiate the plants growing within the frame. Record the code number (obtained from Project A-6) of each kind of plant. Determine and record the number of different kinds of plants. Call this number "A."
4. Choose a different location and repeat Steps 2 and 3. Call the number of different kinds of plants "B."
5. Compare code numbers of the plants found in the first location with those found in the second location. Determine how many kinds of plants are found in both locations. Call this number "C."
6. Determine the Index of Similarity with the following formula:

$$\text{Index of Similarity} = \frac{2(C)}{A + B}$$

7. Determine the Index of Dissimilarity with the following formula:

$$\text{Index of Dissimilarity} = 1 - \frac{2(C)}{A + B}$$

Example:

QUADRAT X CODE NUMBERS	QUADRAT Y CODE NUMBERS
I–11,712,012	I–11,712,012
VIII–46,130,144	III–2,113
IX–150,234,512,653	VIII–36,300,123
IX–157,274,522,364	IX–150,234,512,653
IX–115,234,811,224	IX–175,224,511,342
IX–160,232,312,452	

$$A = 6$$
$$B = 5$$
$$C = 2$$

$$\text{Index of Similarity} = \frac{2(C)}{A + B}$$

$$\text{Index of Similarity} = \frac{4}{11}$$

$$\text{Index of Similarity} = 0.364$$

$$\text{Index of Dissimilarity} = 1 - \frac{2(C)}{A + B}$$

$$\text{Index of Dissimilarity} = 1 - \frac{4}{11}$$

$$\text{Index of Dissimilarity} = 1 - 0.364$$

$$\text{Index of Dissimilarity} = 0.636$$

For Further Investigation

1. Lay out a transect that extends from 10 meters within the edge of a forest to 10 meters into a field. Along this transect, investigate similarity of plant species in at least two quadrats (1 square meter each) within the forest and two quadrats in the field. Account for the different degrees of similarity among the quadrats.

2. Use the methods presented in Project 5-2 to determine the degree of similarity of plants in fields that have not been mowed for different numbers of years.

3. Design and carry out a procedure for determining the Index of Similarity of trees in different forests or in different parts of the same forest.

4. Design and carry out a procedure for determining the Index of Similarity of organisms in water from two different ponds.

Think About This Problem

The Index of Similarity is almost 1 for the plants in two different parks in a particular city.

How might this problem affect you?
What causes the problem?
How could you help solve it?

To find the Index of Similarity of plants in

two quadrats, you first had to determine the total number of plant species in each quadrat. The sum of these two numbers, divided into twice the number of plant species common to the two quadrats gives the Index of Similarity

$$\left(\text{Index of Similarity} = \frac{2(C)}{A + B}\right).$$

Yet, knowing the Index of Similarity does not necessarily give any information about the total number of different species in the quadrats. For example, the similarity index of two quadrats, one in a cornfield and the other on a well-kept lawn, is most likely zero. The plants in one quadrat may be nothing but corn, while the plants in the other may be only one species of grass. Another situation that may show a similarity index of zero would be a quadrat on the edge of a pond and another quadrat high on a hillside. The plants in one habitat are completely different from those in the other habitat. Although this example also shows a similarity index of zero, each of these two quadrats would have a large number of different plant species, instead of only a single species as in the first example.

Farmers, foresters, and gardeners usually try to decrease the number of different species in a habitat. A farmer who is growing soybeans will try to kill off "weeds," which are, in this case, any of the plants that are not soybeans. A forester may plant only spruce trees in a timber lot and eliminate all other trees. Both the farmer and the forester try to permit only one kind of plant to grow in a particular habitat. Such a stand of a single species of plant is called a *monoculture*. Those who raise plants for a living believe they can get the best results by establishing monocultures but this practice is dangerous. For example, the larva of a moth called the spruce budworm (Fig. 5-3) feeds on spruce trees. They kill off huge numbers of spruce trees in the forests where the stands of spruce are monocultures. The Index of Diver-

Fig. 5-3. *A model of spruce budworm eating the leaves of a tree.*
(Courtesy American Museum of Natural History)

sity makes it possible to evaluate several important characteristics of a community. A high Index of Diversity indicates healthy conditions, relative freedom from pollution or other kinds of stress, and stability of the community. Index of Diversity also is an indicator of the degree to which an ecosystem has progressed in the natural succession that tends to take place in the particular habitat. Early stages of succession have a relatively low Index of Diversity. As succession proceeds, the Index of Diversity tends to increase, reaching its maximum at the climax stage of the succession. Because the Index of Diversity can tell you so much about communities, it is very valuable to understand how to calculate it. You will investigate Index of Diversity in Project 5-3.

PROJECT 5-3: Judging the diversity of plants in a field.

You will need: a 2-meter length of string, two plastic jugs (at least 1-liter capacity) with handles, sand, a trowel, a notebook, and a pen.

1. Fill the plastic jugs with sand. Tie one end of the string to the handle of each jug.
2. Choose any point in the field to begin your investigation. Set one jug on the ground, pull the string straight in any direction, and set the second jug on the ground in such a way that the string remains straight. (See Fig. 5-4.)
3. Assign code numbers (using the scheme for differentiating plants—Project A-6 in the Appendix) to each kind of plant along the length of the string. Record the code numbers in your notebook. You may have some difficulty in deciding whether you are looking at one plant or several. Each stem (or in some cases, each leaf) that rises directly from the ground may be considered as belonging to a separate plant.
4. Beginning at one end of the string, observe the first plant; then compare the second plant to the first. Observe whether the second plant is the same as the first. Record your findings in your notebook in the following way. Assign an X to the first plant. If the second plant is the same as the first, write an X for the second plant. If the second plant is different from the first, write an O for the second plant. Continue in this way, using the same symbol (O or X) for the next plant if it is the same as the preceding plant, or using the different symbol (O or X) for the next plant if it is different from the preceding plant.
5. After you have come to the end of the string, move the jugs and string to a new location (which need not be more than a few centimeters away), straighten the string, and repeat Steps 3 and 4. Compare the first plant in this new location with the last plant in Step 4 in order to decide which symbol to assign to the first plant. Continue until you have recorded at least 100 plants.
6. Part of your record may look something like this:

```
XXOOOOOXOXXXOOOXXOXXXXXOOOO
  1     2  34 5   6   789    10
Runs:
Items:  27
```

Each group of similar symbols is called a Run. Each individual plant (or symbol) is called an Item. Denote the runs as shown above.
Count and record the number of runs.
Count and record the number of items.
Count and record the total number of different kinds of plants (called Species).
7. Calculate the Index of Diversity using the following formula:

$$\text{Index of Diversity} = \frac{(\text{number of runs})(\text{number of species})}{(\text{number of items})}$$

In the example given

$$\text{Index of Diversity} = \frac{(10)(4)}{(27)}$$

$$\text{Index of Diversity} = \frac{40}{27}$$

$$\text{Index of Diversity} = 1.48$$

For Further Investigation

1. Find the Index of Diversity of the plants on a lawn. How could the Index of Diversity be used to determine how well kept the lawn is?
2. Find the Index of Diversity for trees in different kinds of forests.
3. Compare the Index of Diversity for plants in a forest to the Index of Diversity for plants in an old field.
4. Find the Index of Diversity for insects that gather around a light at night.
5. Find the Index of Diversity for the organisms that live under a decaying log.
6. Find the Index of Diversity for trees on a city street.

Think About This Problem

Plants are most severely harmed by insects and disease in areas where the Index of Diversity for plants is lowest.

2 EMPTY DETERGENT BOTTLES, TIED WITH STRING, 2 METERS APART

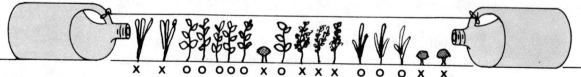

Fig. 5-4. Measuring the Index of Diversity.

How might this problem affect you?
What causes the problem?
How could you help solve it?

Some habitats, such as old fields, have such a rich diversity of organisms that, to avoid confusion, we often must concentrate on one species and ignore all the others. This is what we do when we investigate the distribution of one species in a habitat to try to discover a pattern of distribution and, if possible, the reasons for that pattern. Once we have discovered the reasons for a pattern of distribution of a harmful insect, we may be able to discourage the insect by changing some conditions. For example, cabbage butterfly caterpillars (which harm cabbage) feed on wild mustard. This plant tends to determine the pattern of distribution of the cabbage butterfly. Where wild mustard is plentiful, the cabbage butterflies are abundant. If we destroy the wild mustard in a field, we will decrease the number of cabbage butterfly caterpillars. We might expect similar success in combating other harmful insects if we can determine the reasons for their patterns of distribution.

In Project 5-4 you will investigate the distribution of an insect that is very easy to locate.

PROJECT 5-4: Distribution of the spittlebug.

You will need: a metric tape measure, 21 wooden stakes about 50 centimeters long, a hammer, a quadrat frame (see Project A-4 in the Appendix: Making a quadrat frame), a notebook, and a pen.

1. In the spring, look for the masses of foam that indicate the presence of spittlebug nymphs on low-growing plants. Choose a location in a field where there are numbers of these foam masses (Fig. 5-5).

2. Lay out a transect 20 meters long in the field. Hammer a stake into the ground at every meter mark on the transect. (See Project A-3 in the Appendix: How to lay out a transect.)

3. Set the quadrat frame on the ground with two corners coinciding with the first two stakes on the transect. Count and record the number of spittlebug foam masses within the quadrat. Note that one mass of foam may contain more than one nymph. Since it is time consuming to look for the nymphs inside the foam masses, you will count the number of foam masses rather than the actual number of nymphs.

4. Move the quadrat frame 1 meter further along the transect so that two corners of the quadrat frame coincide with stakes 2 and 3 of the transect. Count and record the number of spittlebug foam masses within the quadrat. Repeat this process until you reach the end of the transect.

Fig. 5-5. Spittlebug foam masses in a field. (Courtesy American Museum of Natural History)

5. Draw a graph of the number of foam masses per quadrat versus the distance along the transect (Fig. 5-6).

6. Describe whatever pattern (or lack of pattern) you discover for the distribution of the spittlebug foam masses. How do you explain your findings?

COMMUNITIES

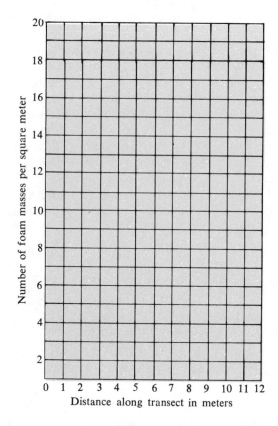

Fig. 5-6. A sample graph to show how the number of spittlebug foam masses per square meter varies along the transect.

For Further Investigation

1. Design and carry out a project similar to Project 5-4 to investigate the distribution of goldenrod stem galls in a field.

2. Design and carry out a project similar to Project 5-4 to investigate the distribution of periwinkle snails on a beach.

3. Count the total number of spittlebug nymphs in 10 foam masses and calculate the average number of nymphs per foam mass. Repeat this procedure for several additional groups of 10 foam masses. Determine how much variation there is in the average number of nymphs per foam mass.

4. Make a map of a field showing the location of each spittlebug foam mass and also the kinds of plants growing in each part of the field.

PROJECT 5-5: Estimating the population of spittlebugs in a field.

You will need: a quadrat frame (see Project A-4 in the Appendix: Making a quadrat frame), equipment for plane table surveying (see Project A-2 in the Appendix: Making a plane table survey), graph paper, a notebook, and a pen.

1. Make a map of the field on graph paper. Set 1 meter in the field equal to the length of a side of a square on the graph paper. (See Project A-2 in the Appendix: Making a plane table survey.)

2. Determine the area of the field in square meters by counting the number of squares (on the graph paper) covered by the map. Count as one-half square meter any square that is partly within and partly outside the outline of the map.

3. Beginning at one edge of the field, throw the quadrat frame several meters in front of you. Count and record the number of spittlebugs within the frame. Repeat this procedure until you have worked your way across the field to the opposite edge.

4. Start at other points on the edge of the field and repeat Step 3 until the area of the field you have sampled is at 5% of the total area of the field.

5. Calculate the average number of spittlebugs per square meter of the field.

6. Multiply the average obtained in Step 5 by

the total number of square meters in the field to obtain the total population of spittlebugs in the field.

For Further Investigation

1. Repeat Project 5-5 in different fields. If the average number of spittlebugs per square meter is significantly different in the different fields, attempt to account for the differences.

2. Remove the spittlebug nymph from its mass of foam. Find out how long it takes for the foam to disappear after the nymph is removed.

3. Repeat Project 5-5 each week for several weeks in the same field to discover how the population changes, if at all.

4. Use methods similar to those in Project 5-5 to estimate the population of dandelions on a lawn.

Think About This Problem

The population of gypsy moths sometimes increases so greatly that their larvae eat all the leaves from some trees (Fig. 5-7).

How might this problem affect you?
What causes the problem?
How could you help solve it?

   Grasshoppers have been competing with man for food for thousands of years. Some of the earliest writings tell of plagues of locusts (grasshoppers) that destroyed entire crops. It is essential to control the grasshopper population, possibly with insecticides, to prevent such disasters. Scientists are constantly searching for more ef-

Fig. 5-7. Gypsy moths showing larva eating a leaf. (Courtesy American Museum of Natural History)

fective methods of combating grasshoppers and other harmful insects. A first step in judging the effectiveness of a grasshopper control method is to measure the population of the grasshoppers and to note how it changes when the control method is used. We cannot employ the quadrat method of estimating population because the grasshopper does not obligingly stay in one place as the spittlebug does. We will use a more suitable method of estimating population in Project 5-6.

PROJECT 5-6: The capture-release-recapture method of estimating population.

You will need: an insect net, a bottle (with brush) of red enamel used for painting model airplanes, four jars with covers, a notebook, and a pen.

1. Use the insect net to catch grasshoppers in a field.

2. As each grasshopper is caught, mark the femurs of its hind legs with red enamel, then put the grasshopper into a jar and cover the jar (Fig. 5-8).

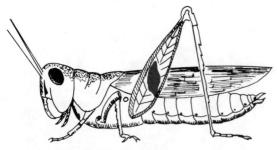

Fig. 5-8. Grasshopper showing red mark on femur.

3. For about 1 hour continue catching grasshoppers and marking their femurs with red enamel. Record the total number of grasshoppers caught and designate the number as A. Release the grasshoppers.

4. One week later, catch grasshoppers in the same field for about ½ hour. These will include some grasshoppers you caught and marked in Steps 2 and 3. Others will be unmarked grasshoppers that were not previously caught. Put the grasshoppers into jars and cover the jars. Do not mark any grasshoppers with enamel.

5. Count and record the number of the grasshoppers captured in Step 4 that have red enamel on their legs. Designate this number as B.

6. Count and record the number of the grasshoppers captured in Step 4 that do not have enamel on their legs. Designate this number as C.

7. Divide B + C by B and multiply the quotient by A. This is an estimate of the total population of grasshoppers in the field.

$$\text{Total population} = A \left(\frac{B + C}{B} \right)$$

8. Explain why the formula in Step 7 gives an estimate of total population.

For Further Investigation

1. Use methods similar to those in Project 5-6 to estimate the total population of tadpoles in a pond.

2. Once each week during late spring, summer, and autumn repeat Steps 3, 4, 5, 6, and 7 of Project 5-6 (after having done the entire project once). Make a graph of the total population of grasshoppers in the field versus time (Fig. 5-9). Explain any changes in population.

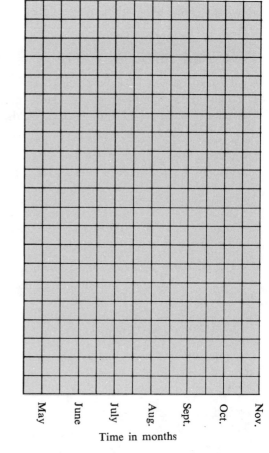

Fig. 5-9. A sample graph showing seasonal variation in the number of grasshoppers in a field.

3. Design and carry out a project similar to Project 5-6 to estimate the total population of ants in a nest.

Think About This Problem

The lower rate at which codfish are caught indicates that the population of codfish has decreased.

How might this problem affect you?
What causes the problem?
How could you help solve it?

Studies show that populations of most species of organisms remain remarkably constant year after year. There are, however, exceptions such as the gypsy moth whose population may experience extremely wide fluctuations. While their numbers in a given locality are generally very small, they can increase explosively to bring about a gypsy moth plague that will destroy trees.

Every species has the potential for a population explosion. This is true even of humans, in spite of their very low reproductive potential which averages about 10 children per female in her lifetime. Many other species have vastly greater reproductive potentials. For example, in one year, one pair of drosophila flies could have millions of offspring.

Uncontrolled population growth of harmful organisms on our small planet can be very frightening indeed. An explosion in rodent population or insect population may lead to epidemics and destruction of vital food crops. Killing off too many predatory animals has lead to overpopulation of herbivores (such as mice, corn borers, and rabbits) which eat much of the cultivated food crops intended for human consumption. That is one important reason why we should observe, measure, and estimate population fluctuations of various organisms. While we are not easily able to investigate the populations of rodents or predatory animals, there is an important area of population that we can look into and possibly do something about.

Goldenrod may be a harmful weed in gardens and fields. It removes much of the available water and nutrients from the soil and thus harms cultivated plants. If goldenrod grows in our community, we can investigate the goldenrod as in Project 5-7 and perhaps find the means to control or eliminate it.

PROJECT 5-7: The reproductive potential of goldenrod.

You will need: a notebook and a pen.

1. Pick an average goldenrod plant in the early autumn, after its parachutelike fruit have ripened, but before any of them have blown away (Fig. 5-10).
2. Remove one fruit head from the plant. Count and record the number of fruit in the fruit head. Designate this number as A.
3. Remove one spray of fruit heads from the plant. Count and record the number of fruit heads in the spray. Designate this number as B.
4. Remove one branch of sprays from the plant. Count and record the number of sprays on the branch. Designate this number as C.
5. Count the number of fruiting branches on the plant. Designate this number as D.
6. To calculate the total number of fruit on the plant, use the following formula:

Total number of fruit $= A \times B \times C \times D$

The total number of fruit may also be considered the reproductive potential of the plant. Each fruit has a single seed. If conditions are perfect, each seed can sprout to form a new goldenrod plant. The reproductive potential of any organism is the

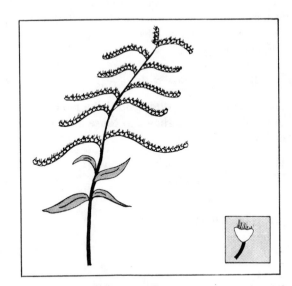

Fig. 5-10. Drawings of goldenrod plant with single fruit head in inset.

number of offspring the organism can produce in its lifetime provided that, as in the case of the goldenrod, the organism is both male and female or else has asexual reproduction (which requires only one "parent"). If both a male parent and a female parent are required, then the number of potential offspring in a lifetime must be divided by 2 to arrive at the reproductive potential.

For Further Investigation

1. To estimate the reproductive potential of drosophila, breed four pairs in a jar containing suitable nutrient material, count the offspring, determine the time, in days, for one generation, then make the necessary calculations.

2. Estimate the reproductive potential of dandelions. Count the number of composite flowers that one dandelion plant produces in one season. After the fruit have ripened, count the number of individual parachutelike fruit on each composite fruiting head. Make the necessary calculations. Another method is to count the dimplelike depressions that indicate where the fruit were attached to the disc that remains on the flower stem after the fruit have been removed.

3. Estimate the reproductive potential of a maple tree. Count the number of samaras (winged maple fruit) in quadrats spaced randomly around the tree. Determine the total area over which the samaras from the tree are scattered. Perform the necessary calculations, observing that each samara contains two seeds.

Think About This Problem

Some herbicides, used to control the populations of weeds, have been shown to cause birth defects in humans.

How might this problem affect you?
What causes the problem?
How could you help solve it?

To maintain the population of the species, each organism need produce, during its life, only one or two offspring that reach maturity and reproduce in turn. An organism that has both sexes in the one individual need have only one such offspring, whereas an organism belonging to a species having one sex per individual must have two offspring that reach maturity. The vast majority of offspring of most organisms do not live to reach maturity. In fact, many eggs are never fertilized, so that the potential offspring do not even begin their own individual existences.

Each organism faces tremendous hazards from the time its life begins, until it finally, if ever, reaches maturity. At any moment the organism's existence may be cut short. Environmental conditions may be unsuitable, it may not be able to withstand competition from other organisms, or it may become food for some other species. If it is to survive to maturity, an organism needs all the help it can get.

Squirrels help oak trees but, of course, they exact a price for their assistance. After acorns have ripened, squirrels become busy, eating as many acorns as they can stuff into themselves, storing quantities of acorns in their winter homes, and burying as many as they can in shallow holes in the ground. During the winter, the squirrels consume the acorns they have stored in their homes. They also dig up and eat some of the buried acorns. It is evident that large numbers of buried acorns are overlooked or never found because in the spring, oak seedlings cover the ground under and around the oak tree.

On one hand, squirrels help the oaks by planting their fruit, the acorns. On the other hand, they exact a price for this service by eating all the acorns they can consume so that not all the acorns grow into oak trees.

In Project 5-8 you will investigate the price the oak pays for the service it receives.

PROJECT 5-8: Squirrels, acorns, and oaks.

You will need: a quadrat frame (see Project A-4 in the Appendix: Making a quadrat frame), a notebook, and a pen.

1. In the autumn locate a mature oak tree that bears many acorns. Just after most of the acorns have dropped from the tree, count the number of acorns in 10 square meters, using the following method. Throw a square meter quadrat frame on the ground 10 times at random in different places within the area under the branches of the tree. Count and record the number of acorns within the quadrat in each location; then find the total. Designate this number as A.

2. In the spring, after oak seedlings have appeared on the ground, repeat the throwing procedure in Step 1 but count empty acorn shells instead of whole acorns in 10 square meters. (The squirrels leave the shells after eating the acorns.) When calculating the number of acorn shells, assume that each piece of shell equals one-half a complete shell. Designate the resulting number as B.

3. Repeat the type of procedure described in Step 1 to count the number of oak seedlings in 10 square meters. Designate this number as C.

4. To calculate the percentage of acorns that the squirrels dug up and ate after burying them, use the following formula:

$$\text{Percentage eaten} = \left(\frac{B}{A}\right)\left(100\right)$$

5. To calculate the percentage of acorns that sprouted, use the formula:

$$\text{Percentage sprouted} = \left(\frac{C}{A}\right)\left(100\right)$$

6. Explain why the number of empty shells found on the ground may be considerably fewer than the number of acorns that dropped from the oak tree in the autumn.

For Further Investigation

1. Watch a squirrel burying an acorn. After he has done so, try to dig up the acorn from the place where he buried it.

2. After the squirrels have buried the acorns in the autumn, dig up 1 square meter of the ground under an oak tree to a depth of about 10 centimeters. Count and record the number of acorns you dig up. Estimate the depth at which the acorns were buried.

3. The following project might be continued for several years. Do not cut the grass or other plants in several square meters of the area in which oak seedlings are growing. Count and record the number of oak seedlings once each month. Make a graph of your results. Explain any decrease in the number of seedlings. Find out where the oak seedlings have the best chance of surviving.

4. Using binoculars, watch squirrels attempting to dig up acorns. Record the success or failure of each digging effort. Estimate the percentage of times that the squirrels are successful in finding acorns when they dig.

Think About This Problem

Some ants protect plant lice and help them thrive. In return the ants get a sweet liquid food from the plant lice.

How might this problem affect you?
What causes the problem?
How could you help solve it?

Instinct guides the squirrel as it buries acorns in the autumn and then attempts to dig them up during the winter and early spring. This adaptive behavior is so valuable to the squirrel and seems so clever that we tend to ascribe more intelligence to the squirrel than it deserves. Instinctive behavior is hereditary and therefore depends on an inherited pattern of genes rather than on intelligence. Instinctive behavior has become effective because those animals that have the most effective behavior have the greatest adaptive advantage. These animals, therefore, have the greatest chance to survive, reproduce, and pass their pattern of behavior on to their offspring along with their genes. Evolution of instinctive behavior can be explained in this way.

We sometimes speak of someone "squirreling away" his money as though his behavior were like that of a squirrel. The person's actions, however, are the result of planning and intelligence; humans have no proven instincts. Intelligence has had the highest adaptive advantage for humans and has displaced instinct. It has made humans superior to all competing animals. Intelligence has enabled humans to develop devices, such as telescopes, airplanes, and weapons that improve upon their natural abilities. Because of their great intelligence, humans are even able to change their environment.

If conditions do not change, instinctive behavior is always advantageous for an animal because only the most effective behavior is inherited by the animal. Such behavior has been proved correct during many generations and possibly for millions of years. On the other hand, there are often harmful side effects when humans use new procedures which they constantly devise with the aid of their marvelous intelligence. Many discoveries that promised to be of greatest benefit to mankind have proved to be disastrous. When DDT was first invented, it seemed as though this poison would free people from the tyranny of insects but now that miracle insecticide must be abandoned because it threatens to plunge numbers of endangered species of animals into extinction. The Aswan Dam was supposed to free Egypt from hunger by providing water to irrigate millions of hectares. Instead, the dam has interfered with agriculture by preventing the annual floods that formerly covered the fields with vitally necessary fertile mud. Moreover, since the completion of the Aswan Dam, there has been a sharp increase in schistosomiasis. This is a debilitating and potentially fatal disease caused by a parasite carried by snails that live in the water. These are only two well-known examples of the failure of supposedly brilliant plans. Before any programs are undertaken to change our environment radically, we must try to foresee what possibly harmful side effects these radical changes might bring to our environment. If, after completion of these programs, there are still undesirable consequences, we must learn from our mistakes and try to improve our procedures. An understanding of ecology can help us avoid error as we plan changes in our technology. Only in this way can we hope to enjoy the marvelous improvements that the other sciences offer us.

CHAPTER 5. COMMUNITIES

1. Why can a large number of different species of organisms live in a rotten log?

2. Why is there a continual, gradual change in the kinds of organisms in a rotting log?

3. The Index of Similarity was calculated for a mixed deciduous forest in Vermont and a desert in Arizona. About what numerical value would you expect the Index of Similarity to have? Explain your answer.

4. Why are monocultures often subject to plagues of insects?

5. What can the Index of Diversity tell you about the health and stability of an ecological community?

6. How does the Index of Diversity change as succession gets closer to the climax stage?

7. Why may the pattern of distribution of a plant species determine the pattern of distribution of an animal species?

8. What are some abiotic environmental factors that can help determine the pattern of distribution of organisms?

9. Why can't the method used for estimating the population of spittlebugs also be used for estimating the population of grasshoppers?

10. Describe a method of estimating the population of tadpoles in a pond.

11. Why is it important to make estimates of the populations of marine fish at least once each year?

12. How may people unintentionally cause a population explosion among field mice?

13. What is reproductive potential?

14. In order to maintain the population of rainbow trout at a constant level, how many offspring of a rainbow trout (on the average) must live to maturity and reproduce?

15. Why is instinctive behavior usually effective?

APPENDIX

PROJECT A-1: How to obtain topographic maps.

To obtain topographic maps of areas in the United States
1. Write to

> Map Information Office
> United States Geological Survey
> Reston, Virginia 22092

Ask for a *Topographic Map Index Circular* for the state in which you are interested and a Geological Survey booklet entitled *Topographic Maps*. These are free.

2. Examine the map included in the *Index Circular*. The entire state is shown, divided into *quadrangles,* each of which is a rectangular map that is identified by the name of a town or important natural feature on the map. Determine which quadrangles you want.

3. Make your order on the map order form you will receive. The 7.5 and 15 minute maps cost 75 cents each. Make your check or money order payable to U.S. Geological Survey.

4. For maps of areas east of the Mississippi River, mail your order and payment to Branch of Distribution, U.S. Geological Survey, 1200 South Eads Street, Arlington, Virginia 22202. For maps of areas west of the Mississippi River, the address is Branch of Distribution, U.S. Geological Survey, Federal Center, Denver, Colorado 80225.

PROJECT A-2: Making a plane table survey.

You will need: a plastic metric ruler, a pot, an electric hot plate, water, a watch, a camera tripod, a metal plate threaded to receive the tripod belt, screws to fit holes in the plate, a screwdriver, a drawing board (about 30 centimeters $\times$ 50 centimeters), a plumb bob and line, and a 20-meter tape measure, two wooden stakes, a hammer, two poles (about 2 meters long and 2 centimeters in diameter), a magnetic compass, plain white paper, masking tape, and a pencil.

1. Pour water into the pot until it is about 5 centimeters deep.
2. Heat the water to boiling over the hot plate.
3. Hold one end of the ruler in the boiling water for 1 minute. Bend the end of the ruler to form a right angle by pressing it at an angle against the bottom of the pot (Fig. A-1).
4. Repeat Step 3 with the other end of the ruler.

APPENDIX

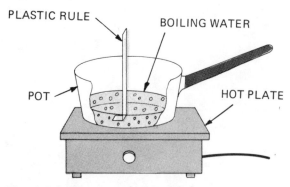

Fig. A-1. How to make an alidade.

You have made an *alidade* from the ruler.

5. Attach the metal plate to the center of the bottom of the drawing board (Fig. A-2).

6. Attach the drawing board to the camera tripod. Attach the plumb bob directly below the center of the drawing board (Fig. A-3).

7. In a field that you wish to survey, lay out a base line, say 50 meters long. Mark each end of the base line by hammering a wooden stake into the ground.

8. Attach a sheet of paper to the drawing board with masking tape.

9. Draw a pencil line 5 cm long on the paper (Fig. A-4).

10. Position the tripod so that one end of the pencil line is directly over the stake at one end of the base line (Point A) as in Fig. A-5. Have a helper hold a pole vertically with one end on the

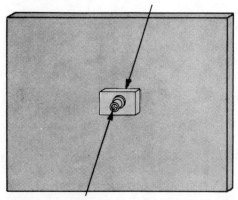

Fig. A-2. Mounting adapter on underside of drawing board.

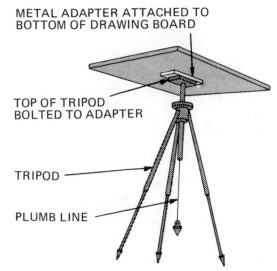

Fig. A-3. Assembling plane table by mounting drawing board to tripod.

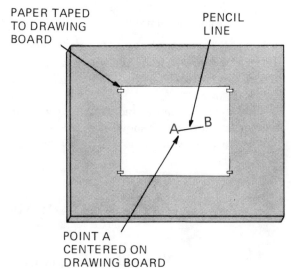

Fig. A-4. Base line AB drawn on a sheet of paper taped to drawing board.

stake at the other end of the base line (Point B) as in Fig. A-6.

11. Set one edge of the alidade on the pencil line. Turn the drawing board until you have lined up the ends of the alidade with the pole. The direction of the pencil line is now the same as that of the base line. Make certain that the setting of the drawing board does not change while it is at Point A.

12. Place the magnetic compass on the paper. Draw a line in the north-south direction. Draw

APPENDIX

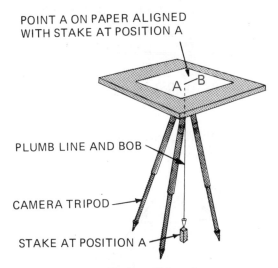

Fig. A-5. Plane table at position A.

an arrowhead on the north end of the line, and label it "N."

13. Place the alidade on the paper so that one edge touches the point that represents Point A. Turn the alidade so that its ends are aligned with a landmark, such as a tree. Draw a pencil line along the edge of the alidade and passing through Point A as in Fig. A-7.

14. Repeat Step 13 for every landmark you wish to represent on your map (Fig. A-8). Label each line to identify the landmark toward which it points.

15. Move the tripod to the other end of the base line (Point B). Move the sheet of paper so that Point B on the map is directly over the center of the drawing board. Fasten the paper to the drawing board with masking tape. Set up the tripod so that Point B on the map is directly over Point B on the base line.

16. Have a helper hold the pole vertically at Point A on the base line, with one end on the stake. Set the alidade on the paper with one edge along the line that represents the base line. Turn the tripod until you have aligned the ends of the alidade with the pole. The direction of the pencil line now is the same as that of the base line. Make certain that the setting of the drawing board does not change while it is at Point B.

17. Repeat Step 13 for each landmark toward which you drew a line from Point A (Fig. A-10 and Fig. A-11). Label each line to identify the landmark toward which it points.

18. The point on your map representing each landmark is at the intersection of the two lines that point toward the landmark. If the lines do not

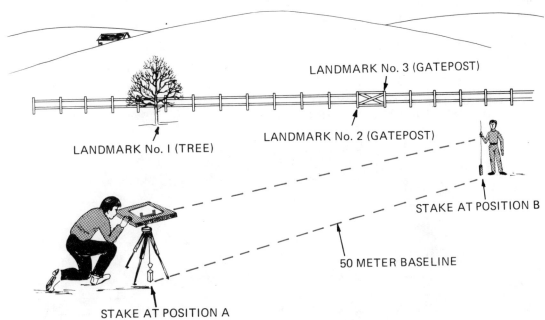

Fig. A-6. Alidade over stake at position A sighted at pole over position B.

Appendix

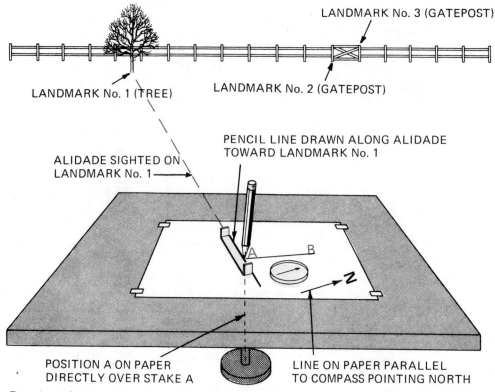

Fig. A-7. Drawing a line on the paper to indicate the direction toward landmark No. 1 (tree) in the field.

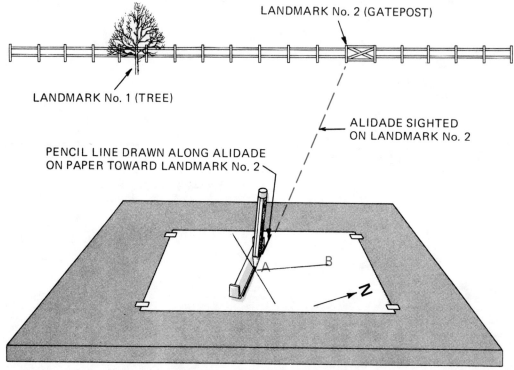

Fig. A-8. Drawing a line on the paper to indicate the direction toward landmark No. 2 (gatepost) in the field.

Appendix

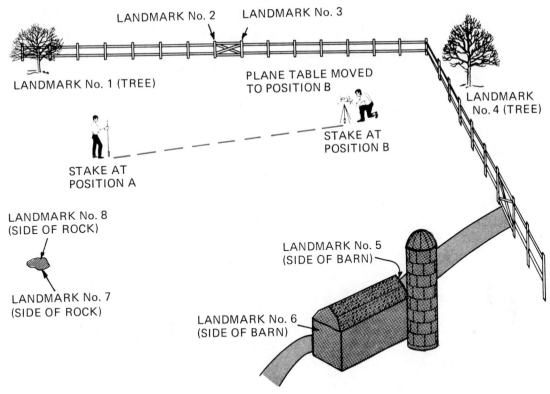

Fig. A-9. Plane table moved from position A to position B.

intersect, extend them in either direction until they do.

19. Using the landmark points as guides, draw the outline of the field (Fig. A-12).

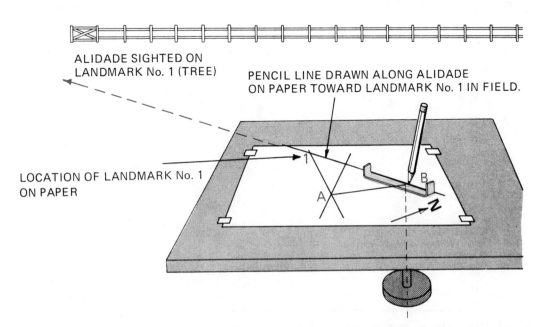

Fig. A-10. Locating the point on the paper corresponding to the location of landmark No. 1 (tree) in the field.

APPENDIX

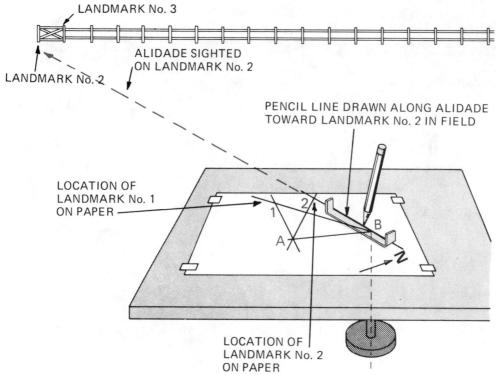

Fig. A-11. Locating the point on the paper corresponding to the location of landmark No. 2 (gatepost) in the field.

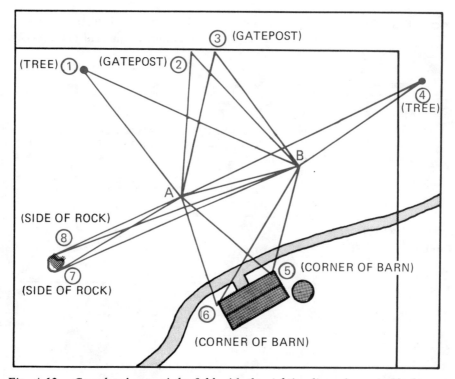

Fig. A-12. Completed map of the field with the sighting lines shown in black.

APPENDIX

PROJECT A-3: How to lay out a transect.

You will need: a metric tape measure, 11 wooden stakes, and a hammer.

1. Hammer a stake into the ground.
2. Using the tape measure, locate the point where you want the transect to end. In this case it is at a distance of 10 meters from the starting point. Hammer a stake into the point where the transect ends.

3. Keep the tape measure extended on the ground from the starting point to the finish point of the transect. Hammer a stake into the ground at each successive meter mark (Fig. A-13). Sight from the stake at the starting point to the stake at the finishing point to align the remaining stakes.

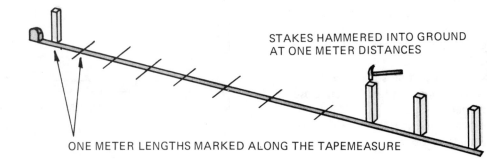

Fig. A-13. *How to lay out a transect.*

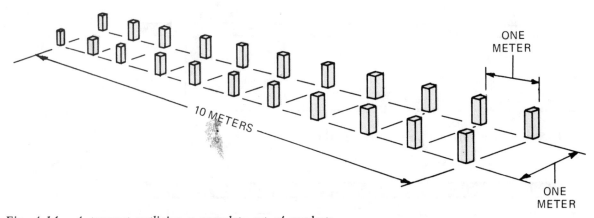

Fig. A-14. *A transect outlining a complete set of quadrats.*

PROJECT A-4: Making a quadrat frame.

You will need: a metric tape measure, two lengths of furring strip (each about 2.5 meters long), a pencil, a car-

APPENDIX

penter's square, a cross-cut saw, four clamps, a drill, a ¼" drill bit, four ¼" × 3" roundhead bolts, four ¼" wing nuts, and a protractor.

1. Measure four 1.1-meter lengths of furring strip. Use the carpenter's square to draw lines perpendicularly across the ends of each length of furring strip (Fig. A-15).
2. Use the cross-cut saw to cut the lengths of furring strip on the lines.
3. Draw a perpendicular line 5 centimeters from one end of each length of furring strip. Draw another line 1 meter from this first line. The second line should be 5 centimeters from the other end of each length of furring strip (Fig. A-16).
4. Draw a line midway between the two sides of the furring strip. Mark a point on this line 2.5 centimeters from the ends of two of the furring strips. Drill a ¼" hole through each of these points (Fig. A-17).
5. Put two lengths of furring strip on a flat surface, parallel to each other and with their closest edges 1 meter apart.
6. Place the two remaining lengths of furring strip with their ends across the ends of the other

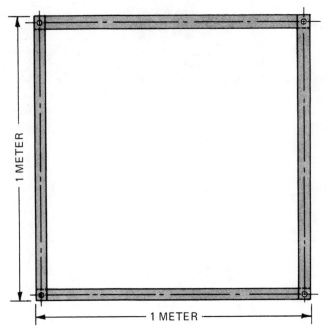

Fig. A-18. Furring strips layed out before bolting together.

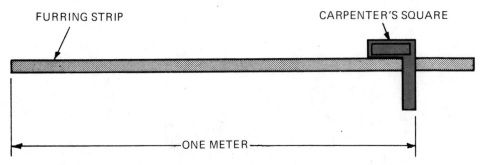

Fig. A-15. Marking furring strips to proper lengths.

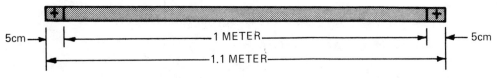

Fig. A-16. Marking center holes at end of furring strip.

Fig. A-17. Holes drilled into furring strips.

furring strips, forming a frame as shown in the illustration. Adjust the lengths of furring strip so that the inner sides of the frame are each 1 meter long (Fig. A-18). The pencil marks will aid in doing this.

7. Use the protractor to set the sides of the frame perpendicular to each other (Fig. A-19). Clamp the corners in position as in Fig. A-20.

8. If necessary turn the frame so that the drilled holes face upward. Using the drilled holes as guides, drill ¼″ holes completely through the corners of the frame.

9. Insert a ¼″ bolt in each hole and turn a wing nut onto each bolt (Fig. A-21). Unfasten the clamps.

NOTE: The quadrat frame may be disassembled to make it easier to carry. When reassembling the quadrat frame, use the protractor to make certain that its sides are perpendicular to each other.

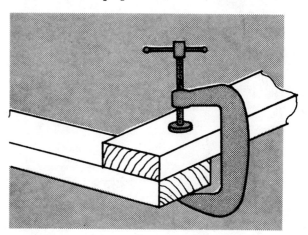

Fig. A-20. Clamping corners of furring strips together at 90° angles.

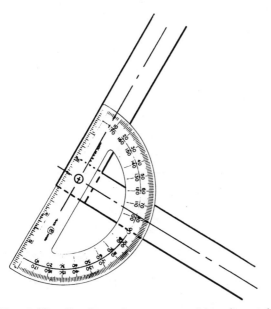

Fig. A-19. Furring strip corners positioned at right angles with protractor.

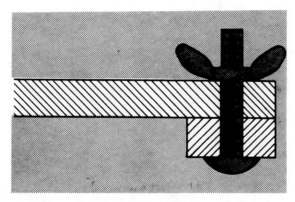

Fig. A-21. Cross section drawing showing how furring strips are bolted together.

APPENDIX

PROJECT A-5: How to measure relative humidity.

You will need: two thermometers, each with a range from about $-10°$ Celsius to about $50°$ Celsius, a woven cotton shoelace, a pair of scissors, a small jar, water, a notebook, and a pen.

1. Cut a piece of the shoelace about 5 centimeters long.
2. Open the piece of shoelace, forming a tube.
3. Slip the tubular piece of shoelace over the bulb of one thermometer. About 3 centimeters of the shoelace should hang past the end of the bulb (Fig. A-22).
4. Wet the shoelace thoroughly. The shoelace can be kept constantly wet by immersing its end in water in a small jar (Fig A-23).
5. Put the two thermometers (the "wet-and-dry-bulb thermometers") where you wish to measure the relative humidity. The place should be shaded.
6. After about 5 minutes, read and record the temperatures on both thermometers. Determine the relative humidity from these temperatures by referring to the following table.
(Table of relative humidity)

RELATIVE HUMIDITY (percentage)

		Dry-bulb temperature (degrees C)																				
		10	11	12	13	14	15	16	17	18	19	20	21	22	23	24	25	26	27	28	29	30
	0.5	94	94	94	95	95	95	95	95	95	95	96	96	96	96	96	96	96	96	96	96	96
	1.0	88	89	89	89	90	90	90	90	91	91	91	91	92	92	92	92	92	92	93	93	93
Difference	1.5	82	83	83	84	85	85	85	86	86	87	87	87	87	88	88	88	88	89	89	89	89
between	2.0	77	78	78	79	79	80	81	81	82	82	83	83	83	84	84	84	85	85	85	86	86
dry-bulb	2.5	71	72	73	74	75	75	76	76	77	78	78	79	80	80	80	81	81	82	82	82	83
and	3.0	66	67	68	69	70	71	71	72	73	74	74	75	76	76	77	77	78	78	79	79	79
wet-bulb	3.5	60	61	63	64	65	66	67	68	69	70	70	71	72	72	73	74	74	75	75	76	76
readings	4.0	55	56	58	59	60	61	63	64	65	65	66	67	68	69	69	70	71	71	72	72	73
	4.5	50	51	53	54	56	57	58	60	61	62	63	64	64	65	66	67	67	68	69	69	70
	5.0	44	46	48	50	51	53	54	55	57	58	59	60	61	62	62	63	64	65	65	66	67
	5.5	39	41	43	45	47	48	50	51	53	54	55	56	57	58	59	60	61	62	62	63	64
	6.0	34	36	39	41	42	44	46	47	49	50	51	53	54	55	56	57	58	58	59	60	61
	6.5	29	32	34	36	38	40	42	43	45	46	48	49	50	52	53	54	54	56	56	57	58
	7.0	24	27	29	32	34	36	38	40	41	43	44	46	47	48	49	50	51	52	53	54	55
	7.5	20	22	25	28	30	32	34	36	38	39	41	42	44	45	46	47	49	50	51	52	52
	8.0	15	18	21	23	26	27	30	32	34	36	37	39	40	42	43	44	46	47	48	49	50
	8.5	10	13	16	19	22	24	26	28	30	32	34	36	37	39	40	41	43	44	45	46	47
	9.0	6	9	12	15	18	20	23	25	27	29	31	32	34	36	37	39	40	41	42	43	44
	9.5		5	8	11	14	16	19	21	23	26	28	29	31	33	34	36	37	38	40	41	42
	10.0			7	10	13	15	18	20	22	24	26	28	30	31	33	34	36	37	38	39
	10.5				6	9	12	14	17	19	21	23	25	27	29	30	32	33	34	36	37
	11.0					6	8	11	14	16	18	20	22	24	26	28	29	31	32	33	35
	11.5						5	8	10	13	15	17	19	21	23	25	26	28	29	31	32
	12.0								7	10	12	14	17	19	20	22	24	26	27	28	30
	12.5									7	9	12	14	16	18	20	21	23	25	26	28
	13.0										6	9	11	13	15	17	19	21	22	24	25
	13.5											6	8	11	13	15	17	18	20	22	23
	14.0												6	8	10	12	14	16	18	19	21
	14.5													6	8	10	12	14	16	17	19
	15.0														5	8	10	12	13	15	17
	16.0																5	7	9	11	13
	17.0																		5	7	9
	18.0																				5
	19.0																				
	20.0																				

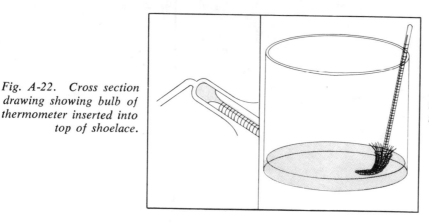

Fig. A-22. Cross section drawing showing bulb of thermometer inserted into top of shoelace.

Fig. A-23. Bulb of thermometer in shoelace tip immersed in water.

PROJECT A-6: Differentiating plants.

Introduction: Many ecological investigations, including some in this book, are concerned with the different kinds of plants that grow in a particular habitat. It is very difficult to identify each plant that you come across by its correct name. Even with years of study, most ecologists cannot identify all of the plants they see. Fortunately, it is not really necessary to identify plants in order to study many of their environmental roles; instead, you can simply differentiate them from each other according to their characteristics. Differentiation is a far simpler process than identification. In this project, the plants are differentiated by giving each a code number, which is different from the code numbers of other kinds of plants. Each digit of a plant's code number represents a particular characteristic of that plant.

It is scarcely possible to make a mistake in differentiating plants, since all that is required is to separate the kinds of plants from each other. It really does not matter, for example, if two people assign a different code number to the same plant. The different kinds of plants will still be separated from each other. To differentiate a plant from others, go through the following steps.

1. Decide on the major group to which the plant belongs. These groups are: I. fungi, II. algae, III. lichens, IV. mosses, V. liverworts, VI. horsetails, VII. club mosses, VIII. ferns, IX. flowering plants and conifers.

The characteristics listed under each of these major groups will help you decide on the group to which a plant belongs. A little experience in differentiating plants will also help greatly.

2. Some important classes of characteristics of plants in each major group are listed after the capital letters (A, B, C, etc.). Decide on the numbered characteristic (after each capital letter) that describes the particular plant. If you cannot make a decision, then the characteristic is indeterminable and the correct number is 0. You may also write a 0 if you do not wish to take the time to decide on some other number for a particular characteristic. Even if the code number you assign to a plant has several 0s in it, you have still coded the plant and have probably separated it from the other kinds of plants.

Write out the selected numbers, in order, to form a code number for the particular plant. For example, a certain kind of mushroom might be assigned the code number I. 31,612,201. This number is assigned because the mushroom is orange, is a typical stemmed mushroom, is between 6 and 10 centimeters in its largest dimension, grows on soil, has its stem approximately under the center of its cap, bears its sporangia on gills, (we could not decide on the texture of the upper surface of its fruiting body—specifically, the cap), and it has an annulus.

I. FUNGI

	1	2	3	4	5	6	7	8	9	0
A. Color	1. White	2. Red	3. Orange	4. Yellow	5. Green	6. Purple or blue	7. Grey	8. Brown or tan	9. Varicolored	0. Indeterminable
B. General shape	1. Stemmed mushroom	2. Approximately spherical	3. Pear-shaped	4. Thin bracket	5. Thick bracket	6. Fingerlike				0. Indeterminable
C. Size of largest dimension in centimeters	1. 0–0.5	2. 0.5–1	3. 1–2	4. 2–4	5. 6–10	6. 6–10	7. 10–15	8. 15–30	9. More than 30 centimeters	0. Indeterminable
D. Substrate in which fungus grows	1. Soil	2. Organic debris or waste	3. Living plant	4. Dead plant	5. Lumber					0. Indeterminable
E. Stem placement	1. No stem	2. Stem under center of cap	3. Stem near edge of cap	4. Stem at edge of cap						0. Indeterminable
F. Structure on which sporangia form	1. Not visible externally	2. Gills	3. Pores	4. "Fringe"	5. Practically smooth surface					0. Indeterminable
G. Texture of upper surfaces of fruiting body	1. Smooth	2. Scaly	3. Shingled	4. Furrowed	5. Pitted					0. Indeterminable
H. Annulus	1. Present	2. Absent								

Definitions: Sporangia—Sport-bearing parts; Annulus—Thin ring around stem.

II. ALGAE

	1	2	3	4	5	6	7	8	9	0
A. Habitat	Salt water	Brackish water	Fresh water pond or lake	Fresh water stream	Soil	Tree or rock				Indeterminable
B. Color	Green	Blue-green	Red	Brown	Purple					Indeterminable
C. General shape	Microscopic	Thin sheet	Feathery	Ribbon-like	Branching ribbon	Leafy	Mossy	Small specks	Flat layer coating a surface	Indeterminable
D. Arrangement of cells	Single cells	Pair of half-cells	Thread of cells joined end to end	Spherical group of cells	Sheet formed of one or more layers of cells					Indeterminable
E. Shape of individual cells	Spherical	Cylindrical	Cubic	Rectangular	Crescent	Triangular	Hexagon	Oval	Irregular	Indeterminable
F. Shape of chloroplasts	Spherical	Oval	Spiral ribbon	Irregular						Indeterminable

III. LICHENS

	1	2	3	4	5	6	7	0
A. Substrate	Soil	Rock	Trees or shrubs					Indeterminable
B. Color	Green	Grey-green	Grey	Brown	Yellow	Pink-red	Varicolored	Indeterminable
C. General shape	Practically flat crusts on substrate	Wavy sheets that stand away from substrate	Hair-like	Branched threads	Upright clubs	Upright clubs each with cup at end	Upright, repeatedly branched	Indeterminable
D. Largest dimensions in centimeters	0–1	1–2	2–5	5–10	10–20	20–30	More than 30	Indeterminable

IV. MOSSES

	1.	2.	3.	4.	5.	6.	7.	8.	9.	0.
A. Leaf edge	1. Smooth	2. Toothed all over	3. Toothed only near tip	4. Tipped with hair-like parts	5. Serrated					0. Indeterminable
B. Ratio of leaf length to leaf width	1. From 1:1 to 2:1	2. From 2:1 to 3:1	3. From 3:1 to 4:1	4. Greater than 4:1						0. Indeterminable
C. Shape of capsule	1. Almost spherical	2. Oval	3. Elongated oval	4. Urn-shaped	5. Pear-shaped					0. Indeterminable
D. Height of gametophyte in centimeters	1. 0.1–0.2	2. 0.2–0.5	3. 0.5–1	4. 1–2	5. 2–4	6. 4–10	7. 10–15	8. More than 15		0. Indeterminable
E. Length of sporophyte in centimeters	1. 0.1–0.2	2. 0.2–0.5	3. 0.5–1	4. 1–2	5. 2–4	6. 4–6	7. More than 6			0. Indeterminable
F. Habitat	1. Submerged in streams	2. Bogs or swamp area	3. Shaded, moist soil	4. Shaded, dry soil	5. Sunny, moist soil	6. Sunny, dry soil	7. Tree trunk	8. Decaying logs	9. Pavement or rocks	0. Indeterminable

Definitions: Capsule—Spore case; Gametophyte—Leafy part; Sporophyte—Capsule and its stalk.

V. LIVERWORTS

These simple relatives of the mosses generally grow flat against the ground, rocks, or bases of trees, although some are aquatic, floating species, and the stems of others may be upright. All the liverworts are small, inconspicuous plants that are easily overlooked.

	1.	2.	3.	4.	5.	6.	7.	8.	0.
A. Growth pattern of the plant	1. Flat, ribbon-like branched or lobed thallus without leaves	2. Leaves growing from branched stem							0. Indeterminable
B. Special characteristics of the thallus if plant is leafless	1. Thallus divided into hexagonal plates with spore in center of each	2. Purple underside and margin of thallus	3. Conspicuous midrib	4. Narrow, much branched	5. Deeply lobed with semi-circular general outline and deeply furrowed surfaces	6. Plant has leaves			0. Indeterminable
C. Shape of leaves if present	1. Almost round	2. Divided into 2 lobes	3. Three large teeth at one end	4. No leaves					0. Indeterminable
D. Habitat	1. Pond, swamp or marsh	2. Damp soil	3. Rocks	4. Tree trunks					0. Indeterminable
E. Width in centimeters	1. 0.1–0.2	2. 0.2–0.3	3. 0.3–0.4	4. 0.4–0.6	5. 0.6–0.8	6. 0.8–1	7. 1.0–1.5	8. More than 1.5	0. Indeterminable
F. Length in centimeters	1. 0–0.5	2. 0.5–1	3. 1–2	4. 2–4	5. 4–6	6. 6–8	7. 8–12	8. More than 12	0. Indeterminable

VI. HORSETAILS

These plants have stems divided into segments by raised horizontal ridges, which make them look somewhat like miniature bamboo. The stems of most species also have vertical ridges.

A. Growth pattern	1. Curled stems	2. Straight stems without branches	3. Straight stems with whorls of undivided branches growing from most joints	4. Straight main stems with whorls of repeatedly divided branches growing from most joints	0. Indeterminable
B. Texture of stem	1. Smooth	2. Flat, vertical ridges	3. Rough, vertical ridges		0. Indeterminable
C. Structure of stem	1. Hollow	2. Solid			0. Indeterminable
D. Habitat	1. Forest	2. Field	3. Sandy areas	4. Near marshes, swamps, ponds, or streams	0. Indeterminable
E. Height in centimeters	1. 0–10	2. 10–30	3. 30–60	4. 60–100	5. 100–200	0. Indeterminable

VII. CLUBMOSSES

These are evergreen vines that are somewhat moss-like, but much coarser than true mosses. Upright branches grow from creeping horizontal mainstems.

A. Growth pattern of upright branches	1. Unbranched	2. Sparsely branched	3. Much branched with almost straight branches	4. Much branched with curved, drooping branches	0. Indeterminable
B. Shape of leaves	1. Scale-like	2. Long-oval	3. Hair-tipped	4. Sharp-tipped	0. Indeterminable
C. Characteristics of spore bearing organs (usually cones)	1. No cones	2. Spore cases located in broadened tips of upright branches	3. Single cone grows directly from tip of almost every upright branch	4. Clusters of cones on almost leafless stems rise from upright branches	0. Indeterminable
D. Relative thickness of horizontal stems	1. Considerably thinner than the upright branches	2. About as thick as the upright branches	3. Considerably thicker than the upright branches		0. Indeterminable
E. Leafiness of horizontal stems	1. Leafy	2. Almost leafless			0. Indeterminable
F. Habitat	1. Bogs, swamps, or marshes	2. Forests			0. Indeterminable
G. Height in centimeters	1. 0–10	2. 10–20	3. 20–30	4. More than 30	0. Indeterminable

VIII. FERNS

	1	2	3	4	5	6	7	8	9	0
A. Dissection (cutting) or lobation	1. Entire (not cut or lobed)	2. Shallowly lobed (less than halfway to midrib)	3. Deeply lobed (more than halfway to midrib)	4. Once cut (to midrib)	5. More than once cut					0. Indeterminable
B. Shape of blade	1. Grass-like	2. Oval	3. Broadest at base	4. Broadest at middle	5. Broadest above middle	6. Broadest near but not at base	7. Blade palmately branched	8. Frond branched forming 3 parts		0. Indeterminable
C. Margin of pinnae, pinnules, or blade	1. Entire	2. Toothed								0. Indeterminable
D. Shape of pinnae	1. Almost round	2. Triangular	3. Almost oval	4. With auricle (ear-shaped part) next to rachis (stem of blade)	5. No pinnae					0. Indeterminable
E. Pinnules (sections of pinna)	1. First pinnule (next to rachis) on lowest pinna longest	2. Second pinnule (from rachis) on lowest pinna longest	3. Longest pinnule other than first or second	4. No pinnules						0. Indeterminable
F. Color of Stipe (base of rachis) and rachis	1. Entirely green	2. Entirely dark brown	3. Entirely black	4. Stipe dark brown at base	5. Stipe dark brown up to blade	6. Stipe and most of rachis dark brown				0. Indeterminable
G. Scales and hair	1. Blade hairy underneath	2. Stipe covered with hairs	3. Stipe and rachis covered with hair	4. Stipe, rachis, and veins covered with hairs	5. Blade scaly underneath	6. Stipe covered with scales	7. Stipe and rachis covered with scales	8. Stipe, rachis, and blade scaly	9. No hairs or scales, or only at the extreme base	0. Indeterminable
H. Arrangement of spore cases	1. Under the rolled edges of blade	2. In two rows between edges and midveins	3. Scattered on underside of blade	4. In open cups on edges of blade	5. In a spike on a stalk	6. Spherical branches of a stalk rising from ground	7. Spherical, on branches of a stalk rising from blade	8. Spherical, on several middle pinnae of some blades		0. Indeterminable
I. Length in centimeters	1. 0–0.5	2. 0.5–1	3. 1–5	4. 5–10	5. 10–30	6. 30–60	7. 60–100	8. 100–200	9. 200–400	0. Indeterminable
J. Width in centimeters	1. 0–0.5	2. 0.5–1	3. 1–5	4. 5–10	5. 10–30	6. 30–60	7. 60–100	8. 100–200		0. Indeterminable
K. Habitat	1. Fields	2. Moist forests	3. Near standing or flowing water, bogs	4. Wet rock ledges (not limestone)	5. Wet limestone ledges	6. Dry limestone ledges	7. Moss-covered rocks	8. Hillsides	9. On trees	0. Indeterminable

IX. FLOWERING PLANTS AND CONIFERS

	1	2	3	4	5	6	7	8	9	0
A. Type of growth	1. Plants without woody stems	2. Shrubs	3. Trees	4. Vines with woody stems						0. Indeterminable
B. Flower color	1. White	2. Pink	3. Red	4. Orange	5. Yellow	6. Green	7. Blue	8. Purple	9. Varicolored	0. Indeterminable
C. Number of petals	1. 0	2. 1 (usually tubular)	3. 3	4. 4	5. 5	6. 6	7. 7 or more			0. Indeterminable
D. Leaf groupings	1. No leaves	2. Leaves growing separately on stem	3. Leaves growing in pairs on stem	4. Leaves growing in groups of 3 on stem	5. Leaves growing in groups of 5 on stem	6. Leaves growing in groups of more than 5 on stem				0. Indeterminable
E. Leaf placement	1. No leaves	2. Leaves opposite each other along stem	3. Leaves alternate along stem	4. Leaves in whorls along stem	5. Leaves in two densely packed rows on opposite sides of stem	6. Leaves densely scattered along stem	7. Leaves only at base of plant			0. Indeterminable
F. Venation	1. No veins visible	2. Main veins parallel (monocot)	3. Palmate veining	4. Pinnate veining						0. Indeterminable
G. Leaf shape	1. Scale-like	2. Needle	3. Long and narrow	4. Almost round	5. Oval	6. Wider at tip than at base	7. Wider at base than at tip	8. Compound with 3 leaflets	9. Compound with more than 3 leaflets	0. Indeterminable
H. Lobation or dissection	1. Not dissected	2. Dissected less than halfway to midrib	3. Dissected more than halfway to midrib	4. Palmately compound	5. Pinnately compound	6. More than one compound				0. Indeterminable
I. Leaf edge	1. Entire	2. Toothed simply	3. Compoundly toothed (each tooth has secondary teeth)	4. Wavy	5. Lobes ending in points					0. Indeterminable
J. Average height of plant in centimeters	1. 0–2	2. 2–5	3. 5–10	4. 10–20	5. 20–40	6. 40–100	7. 100–200	8. 200–500	9. More than 500	0. Indeterminable
K. Average leaf length in centimeters	1. 0–1	2. 1–2	3. 2–4	4. 4–6	5. 6–8	6. 8–10	7. 10–15	8. 15–30	9. More than 30	0. Indeterminable
L. Average leaf width in centimeters	1. 0–0.1	2. 0.1–0.5	3. 0.5–1	4. 1–2	5. 2–4	6. 4–6	7. 6–10	8. 10–30	9. More than 30	0. Indeterminable

I. Fungi

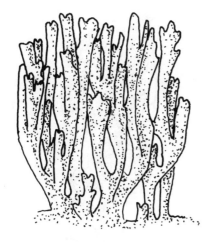

Clavicorona
I 86,641,512

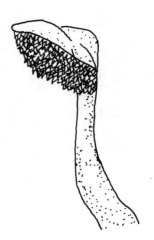

Pine cone fungus
I 21,714,422

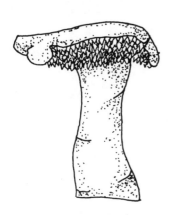

Repand hydnum
I 81,713,412

Multizoned polystichus
I 84,641,342

Appendix

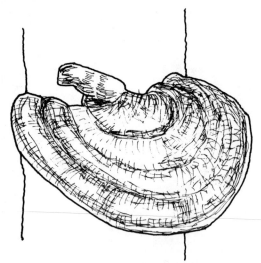

Ganoderma tsugae
I 15,844,312

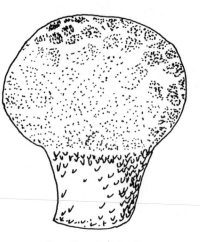

Pearshaped puffball
I 83,511,112

Lepiota rachodes
I 71,822,231

Common morel
I 91,712,052

Giant puffball
I 12,911,112

122

II. Algae

Sea lettuce
II 112,005

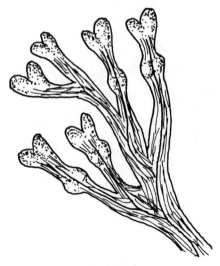

Rockweed
II 145,000

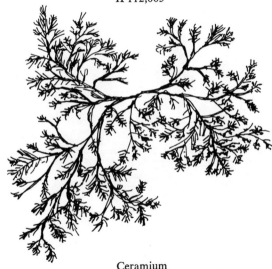

Ceramium
II 133,000

Mermaid's hair
II 127,000

Kelp
II 144,000

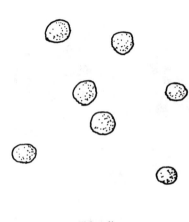

Chlorella
II 311,110

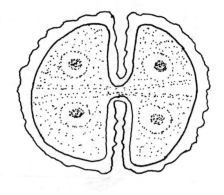

Cosmarium
II 311,290

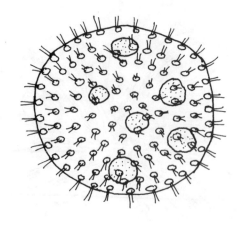

Volvox
II 311,410

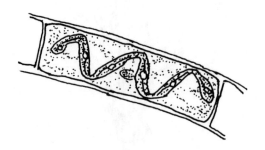

Spirogyra
II 311,323

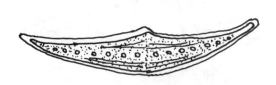

Closterium
II 311,250

III. Lichens

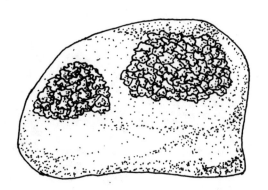

Map lichen
III 2,510

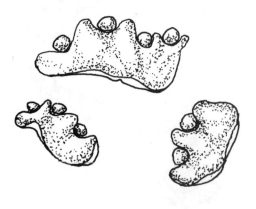

Fan lichen
III 1,222

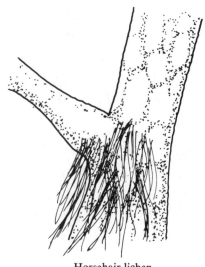

Horsehair lichen
III 3,434

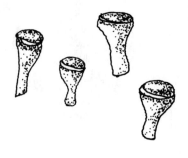

Pixie cup
III 2,263

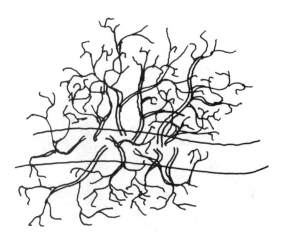

Golden lichen
III 3,544

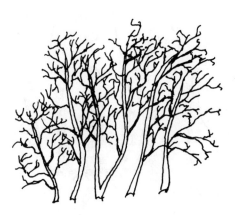

Reindeer lichen
III 1,376

British soldiers
III 1,753

IV. Mosses

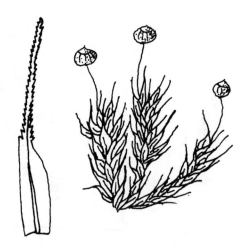

Apple moss
IV 241,654

Urn moss
IV 124,333

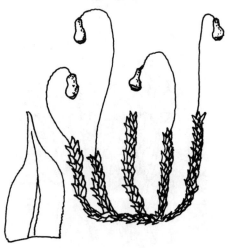

Silver moss
IV 115,336

Shaggy moss
IV 512,858

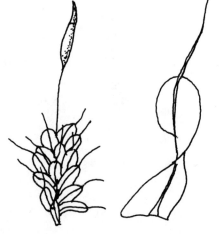

Wall moss
IV 433,359

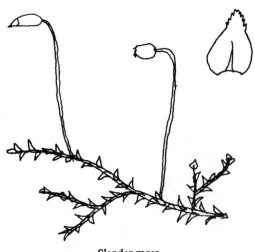

Slender moss
IV 312,360

V. Liverworts

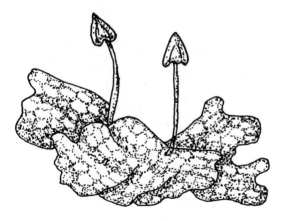

Great scented hornwort
V 114,277

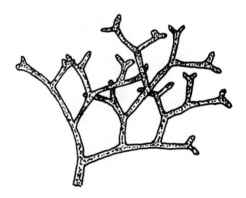

Slender riccia
V 144,115

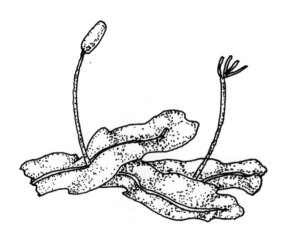

Common scale moss
V 134,227

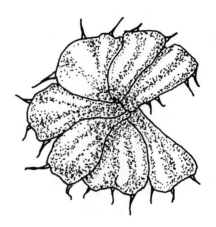

Purple-fringed riccia
V 154,173

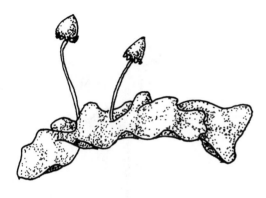

Asterella
V 124,224

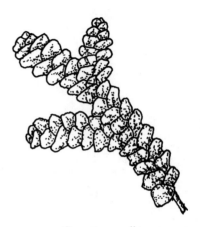

Common porella
V 261,446

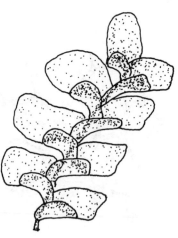

Common scapania
V 262,277

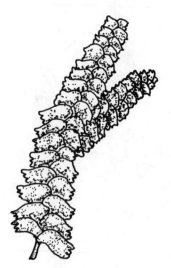

Three-lobed bazzania
V 263,277

VI. Horsetails

Dwarf horsetail
VI 13,212

Field horsetail
VI 33,133

Smooth horsetail
VI 21,145

Wood horsetail
VI 43,113

VII. Club Mosses

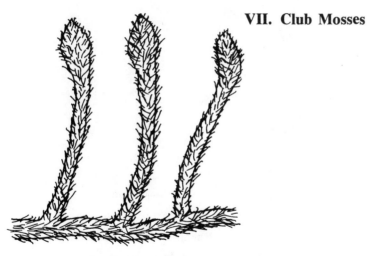

Foxtail club moss
VII 1,122,113

Tree club moss
VII 3,142,223

Shining club moss
VII 2,412,122

Running ground pine
VII 4,143,223

Stiff club moss
VII 2,231,122

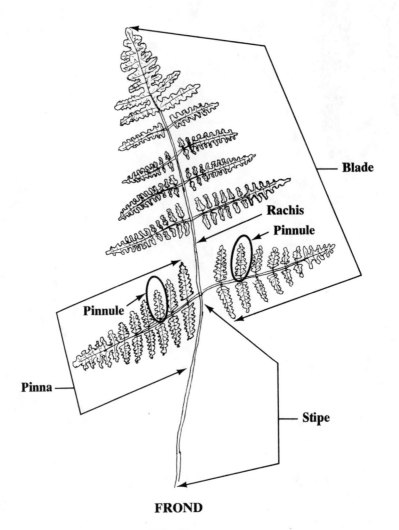

FROND

VIII. Ferns

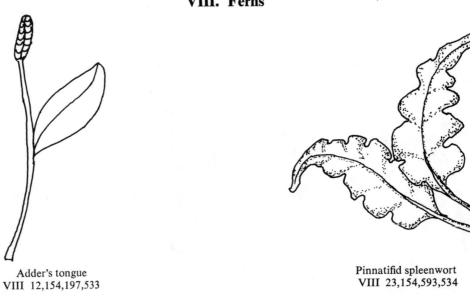

Adder's tongue
VIII 12,154,197,533

Pinnatifid spleenwort
VIII 23,154,593,534

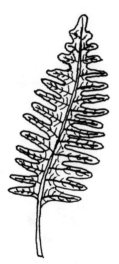

Common polypody
VIII 36,154,192,642

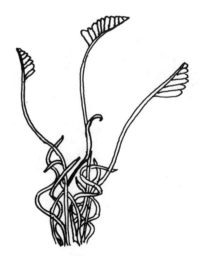

Curly grass
VIII 11,154,196,413

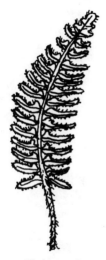

Christmas fern
VIII 44,244,472,852

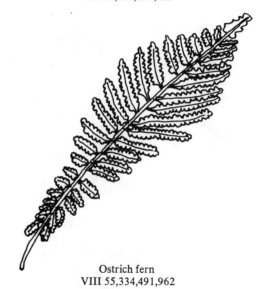

Ostrich fern
VIII 55,334,491,962

Bracken
VIII 58,323,091,881

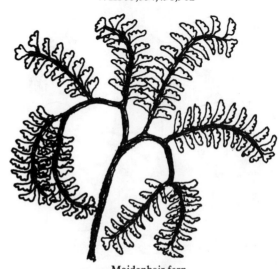

Maidenhair fern
VIII 57,334,391,652

IX. Flowering Plants and Conifers

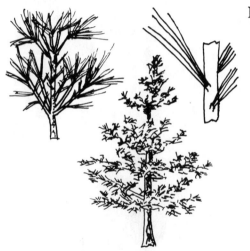

White pine
IX 300,561,211,972

Solomon's seal
IX 162,232,511,675

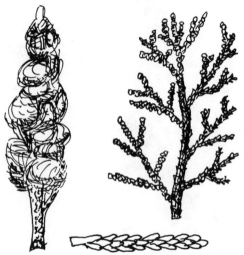

Red cedar (eastern)
IX 300,221,111,912

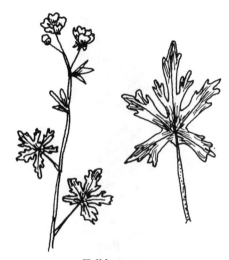

Tall buttercup
IX 155,233,931,657

Black willow
IX 301,234,312,974

Wistaria
IX 474,234,952,977

Wood lily
IX 196,242,312,654

Cinquefoil
IX 155,234,942,446

ECOLOGY TEXTBOOKS

Andrews, William A., **Contours: Studies of the Environment—Soil Ecology,** Prentice Hall, N.J., Prentice-Hall, 1974.

Andrews, William A., **Contours: Studies of the Environment—Terrestrial Ecology,** Prentice-Hall, Englewood Cliffs, N.J., 1974.

Blaustein, Elliott H. *et al.*, **Your Environment and You,** Oceana, Dobbs Ferry, N.Y., 1974.

Buchsbaum, R., and Buchsbaum, M., **Basic Ecology,** Boxwood Press, Pittsburgh, Pennsylvania, 1957.

Dansereau, Pierre, **Biogeography**, Ronald Press Company, New York, 1957.

Dice, Lee R., **Natural Communities**, The University of Michigan Press, Ann Arbor, Michigan, 1968.

Odum, E., **Ecology,** Holt, Rinehart, New York, 1969.

Watt, Kenneth E. F., **Principles of Environmental Science**, McGraw-Hill Book Company, New York, 1973.

Human Ecology

Busch, Phyllis, **People and Their Environment: The Urban Environment**, Doubleday, Garden City, N.Y., 1971.

Caldwell, William A., **How to Save Urban America**, New American Library, Inc., Signet Books, New York, 1973.

Ehrlich, Paul, and Ehrlich, Anne, **Population, Resources, Environment**, Freeman, San Francisco, 1970.

McHarg, I., **Design with Nature,** Natural History Press, Garden City, N.Y., 1969.

Meadows, Donella H., *et al.*, **Limits to Growth**, Universe, New York, 1972.

Reilly, William K., **The Use of Land— A Task Force Report Sponsored By the Rockefeller Brothers Fund**, Thomas Y. Crowell Company, New York, 1973.

Roth, Gabriel, **Paying for Roads**, Penguin Books, Baltimore, 1967.

Wagner, Richard H., **Environment and Man,** W. W. Norton, Inc., New York, 1974.

Whyte, William H., **The Last Landscape**, Doubleday, Garden City., N.Y., 1968.

Specialized Topics

Farb, Peter, **The Forest**, Time, Inc., New York, 1961.

Jackson, R. M., and Raw, F., **Life in the Soil**, St. Martin's Press, New York, 1966.

Niering, W. A., **Life of the Marsh**, McGraw-Hill, New York, 1967.

Storer, John H., **The Web of Life**, New American Library, New York, 1956.

Usinger, R. L., **Life of the Rivers and Streams**, McGraw-Hill, New York, 1967.

Wynne-Edwards, V. C., "**Population Control in Animals**," *Scientific American*, 211 (June): 68–74, 1964.

"**The Cities**," *Scientific American*, 213 (September): 1965.

Environmental Techniques and Activities

Graybill, Roy, and Walters, Sandra, **Environmental Education on an Urban School Site**, National Park Service, Seattle, Washington, n.d.

Hillcourt, William, **Fieldbook of Nature Activities**, G. P. Putnam's Sons, New York, 1950.

Johnson, G., and Bleifeld, M.,
Hunting with the Microscope, Arco,
New York, 1975.

Pettit, Ted, **The Book of Nature Hobbies**, Didier,
New York, 1947.

Field Guides

Borror, D. J. and White, R. E.,
Insects, A Field Guide, Houghton Mifflin,
Boston, 1970.

Chu, H. F., **How to Know the Immature Insects,**
Wm. C. Brown, Dubuque, Iowa, 1949.

Cobb, Boughton, **A Field Guide to Ferns,**
Houghton Mifflin, Boston, 1974.

Conant, Roger, **A Field Guide to Reptiles and Amphibians of Eastern North America,**
Houghton Mifflin, Boston, 1958.

Conrad, Henry S., **How to Know the Mosses and Liverworts**, Wm. C. Brown,
Dubuque, Iowa, 1956.

Cuthbert, Mable J., **How to Know the Fall Flowers**, Wm. C. Brown,
Dubuque, Iowa, 1948.

Cuthbert, Mable J., **Know How to Know the Spring Flowers**, Wm. C. Brown,
Dubuque, Iowa, 1949.

Ehrlich, Paul R., **How to Know the Butterflies,**
Wm. C. Brown, Dubuque, Iowa, 1961.

Jahn, T. L., **How to Know the Protozoa,**
Wm. C. Brown, Dubuque, Iowa, 1949.

Jaques, H. E., **How to Know the Insects**,
Wm. C. Brown, Dubuque, Iowa, 1949.

Jaques, H. E., **How to Know the Plant Families**,
Wm. C. Brown, Dubuque, Iowa, 1947.

Kaston, B. J., and Kaston E., **How to Know the Spiders,** Wm. C. Brown, Dubuque, Iowa, 1947.

Lutz, Frank E., **Field Book of Insects**,
Putnam, New York, 1935.

McKenney, Margaret, and Peterson, R. T.,
A Field Guide to Wildflowers, Houghton Mifflin,
Boston, 1974.

Oliver, James A., **North American Amphibians and Reptiles**, D. Van Nostrand, New York, 1955.

Peterson, Roger Tory, **A Field Guide to the Birds**,
Houghton Mifflin, Boston, 1947.

Petrides, George A., **A Field Guide to Trees and Shrubs**, Houghton Mifflin, Boston, 1958.

Prescott, G. W., **How to Know the Fresh Water Algae**, Wm. C. Brown,
Dubuque, Iowa, 1964.

Smith, Alexander H., **The Mushroom Hunter's Field Guide**, University of Michigan Press,
Ann Arbor, Michigan, 1958.

Watts, May T., **Master Tree Finder**,
Nature Study Guide, Berkeley, California, 1963.

Watts, May T., and Watts, Tom,
Winter Tree Finder, Nature Study Guild,
Berkeley, California, 1970.

Zim, Herbert S., and Smith, H.,
Reptiles and Amphibians, Simon and Schuster,
New York, 1956.

Zim, Herbert S., and Coltram, Clarence,
Insects, Simon and Schuster, New York, 1951.

Zim, Herbert S., and Shuttleworth, Floyd S.,
Non-Flowering Plants, Golden Press,
New York, 1967.

INDEX

Abiotic factors, 27, 37, 52, 99
Africa, 53
Aldehydes, 27
Alders, 73
Algae, 27, 35, 60, 65
Alidade, use of, 79, 80
Animals, 51
    increase in mass and relation to food intake, 60–61
    See also individual species.
Annual ring quotients, 50–51
Ants, 67, 97
Asparagus, "blanched," 40
Aswan Dam, 98
Automobiles, 25–30
    exhaust as pollutants, 27–28, 29, 30, 35
    mass transportation as alternative, 30
    parking facilities, 25–27
    speed of, and traffic volume, 29–30, 35

Bacteria, 74, 75, 76, 84
Balance of nature, 65, 76, 82
    in paramecia and didinia, 76–77, 78
Bass, 45
Bean culture, 76
Beavers, 78
Bees, 72–73
    territoriality of, 81–82
    value as pollinators, 73, 84
Biotic factors, 37
Birds. *See* individual species.
Boating, 65
Brook trout, 45

Cacti, 54
California, house slides in, 20
Cancer, 65
Capillarity, 53
Carbon dioxide, 56, 57
Carbon monoxide, 27
Carp, 43, 45
Catfish, 43, 45
Cattail plants, 73

Cellars, wet or flooded, 19, 25
Chloroplasts, 40
Cities
    advantages of, 30–31
    open spaces in, 33
Cockroaches, 67, 68, 70, 84
Communities, 83, 84, 85–99
    in field, 87–88, 89–91
    in rotten log, 85–87, 99
    See also individual species.
Commuting, 31
Crayfish, 20
Critical land, 16, 17–22
    as building sites, 16, 17, 20
    laws protecting, 20
    zoning of, 19
    See also Real estate development.

DDT, 98
Deer, 20
Deserts, 49, 51, 53, 54–55
Didinia, 76–77, 78
Drosophila flies. *See* Fruit flies.
Dry wells, 25
Ducks, 20

Earthworms, 67
Ecological communities. *See* Communities.
Ecological succession. *See* Succession, ecological.
Ecology, 11
Egypt, 98
Electrical energy use, 31–33, 35
    brownouts and blackouts, 33
Energy
    electrical, 31–33, 35
    solar, 56
    transfer of, 51, 53–66
Environment
    assessment of, 15–16
    disruption of, 16
    factors in, 37–52
    niches in, 65, 66–73, 82–83, 84, 87
Erosion, 16, 19, 25

INDEX

Etiolation, 38–39, 52
Eutrophication, 65
Evaporation, 53
Evolution, 97
Exterminators, 67

Farmland, 16
Fertilizers, 65, 66
Fields, 87–88, 89–91
Fish, 52, 99
   and pollution, 58
   See also individual species.
Fishing, 27, 65
Floodplains, 16, 19, 20, 21, 22, 35
Floods and flooding, 16, 19, 20, 21, 22, 24, 35, 98
Forests, destruction of, 16
Fruit flies, 68–69, 84, 94, 96

Gall insects, 70–72, 84
Garfish, 43, 45
Glucose, 57
Goldfish, 45, 57
Grasshoppers, 93, 99
Gypsy moths, 93, 94–95

Habitat, 84
   destruction of, 20, 82–83
   See also Niches, environmental.
Herbicides, 96
   See also Insecticides.
Herbivores, 60, 66
Holding ponds, 24–25
Honeybees. *See* Bees.
Houseflies, 70
Humidity, relative, 55–56
Hydrocarbons, 27
Hydrogen, 57

Index of Dissimilarity, 87, 88
Index of Diversity, 89–90, 99
Index of Similarity, 87, 88, 89, 99
Insecticides, 93, 98
   See also Herbicides.
Insects
   distribution of, 91–93
   food preferences of, 67–70, 84
   harmful, 67, 93
   population estimates, 92–94
   See also individual species.
Instinctive behavior, 97, 98, 99

Irrigation, 55
Israel, irrigation in, 55

Lakes
   eutrophication of, 60, 65
   siltation of, 16
Land development. *See* Real estate development.
Lumber companies, 75
Lynx, 76, 84

Man
   as adapter of environment, 34, 37, 43, 65, 97, 98, 99
   birth defects in, 96
   reproductive potential of, 95
Maps
   floodplain, 17, 18, 19
   topographic, 17
Marshes, 25, 35
Mass transportation, 30
Microclimates, 48, 52
Monoculture, danger of, 89, 99
Muskrats, 20

Natural cycles, 51, 53–66
Niches, environmental, 65, 66–73, 82–83, 84, 87
Nitrates, 27, 65
Nitric acid, 27
Nitrogen oxides, 27

Oaks, 96–97
Open spaces, 33–34, 35
Oxygen, 37, 45, 52, 57, 58
   produced during photosynthesis, 58–59
Ozone, 27

Parking facilities
   environmental damage caused by, 27–28
   evaluating efficiency of, 25–27
Parks. *See* Open spaces.
Peracyl nitrate, 27
Photosynthesis, 37, 40, 42–43, 56, 58–59, 60, 66
Plant lice, 97
Plants, 37, 51
   composition of 61–64, 66
   dry mass of, 39, 52
   herbaceous, 47
   increasing damage to, 70
   sprouting of, 46–48
   transplanting of, 54
   See also individual species.

INDEX

Pollutants and pollution
   in automobile exhaust, 27–28
   noise, 46
   thermal, 43, 45, 52
Population density, 30–31, 35
   and energy requirements, 31
   and open spaces, 33–34
Population measure and control, 92–96
Predator/prey relationship. *See* Balance of nature.
Pupation, 45

Quadrat (frame), 40, 42, 87, 88, 91, 92

Rabbits, 76, 84
Rainfall, 19
   effect on tree growth, 49–51
   importance of, 49, 53
   quotients, 50
   variations in, 50–51
Real estate development
   as cause of erosion, 20, 22
   as cause of floods, 22, 25, 35
   on critical land, 16, 17–22
   and destruction of habitat, 82–83
   See also Runoff.
Red foxes, 78, 84
Redwoods, 75
Reproductive potential, 94, 95–96, 99
Rivers, 22, 35
   siltation of, 16
Robins, 67, 78–79, 84
Root hairs, 53
Runoff
   from developed land, 22, 23–25, 35
   and parking lots, 25

Sahel (region of Africa), 51
Sandhill cranes, 79
Schistosomiasis, 98
Science, value of, 11
Seasons, cause of, 43
Sewage problems, 19, 58
Sewers
   catch basins in, 23
   and holding ponds, 24–25
   sanitary, 23
   storm, 22, 23, 24, 25
Slope-measuring of land, 18, 35

Soil
   capacity for holding water, 25, 53
   mineral composition of, 64–65
Soil erosion. *See* Erosion.
Spiders, 70
Spittlebugs
   distribution of, 91
   population estimates of, 92–93, 99
Spruce budworm, 89
Squirrels, 96–97
Streams, 25
   calculation of cross-sectional area, 24
   destruction of, 20
   water velocity of, 24, 35
Succession, ecological, 73–76, 84, 87, 99
Sulfur oxides, 27
Sunlight, 37
   effect on plant growth, 37–38, 40–43, 52
   intensity of, 40–42, 52
   See also Etiolation.
Swamps, 25
   draining of, 20, 35, 82
Swimming, 27, 65

Temperature, 43–49
   difference, on slopes, 48–49
   effect on fish, 43–46
   effect on mealworm pupae, 45
   effect on sprouting, 46–48, 52
Termites, 68, 69
Territoriality, 65, 76–82
   in beavers, 78
   in bees, 81–82
   in red foxes, 78, 84
   in robins, 78–79, 84
   in wolves, 78
   in yellow jacket hornets, 79–81
Trace elements, 65, 66
Transect, 40
Transpiration, 53–54, 66
Trees
   climax, 75, 76, 99
   conifers, 54
   deciduous, 54
   pioneer, 76, 84
   rainfall and, 49–51
   succession in, 76
Trout, 43, 45

United States, irrigation in, 55
U.S. Geological Survey, 17

Vegetarian diets, 60, 61, 66

Water
   algae as pollutants, 27
   recreational, 27
Water cycle, 51, 52
Watersheds, 25

Water supply, decrease of, 16, 20
   See also Water table.
Water table
   cross section, 21
   lowering of, 20
Willows, 73
Wolves, 78

Yeasts, 74–75, 76, 84
Yellow jacket hornets, 79–81